Neutron and X-ray Reflectometry

Emerging phenomena at heterostructure interfaces

Online at: https://doi.org/10.1088/978-0-7503-4695-5

Neutron and X-ray Reflectometry

Emerging phenomena at heterostructure interfaces

Saibal Basu
Ex. Solid State Physics Division, Bhabha Atomic Research Centre, Homi Bhabha National Institute, Mumbai 400094, India

Surendra Singh
Solid State Physics Division, Bhabha Atomic Research Centre, Mumbai 400085, India
and
Homi Bhabha National Institute, Anushakatinagar, Mumbai 400094, India

IOP Publishing, Bristol, UK

ISBN 978-0-7503-4695-5 (ebook)
ISBN 978-0-7503-4693-1 (print)
ISBN 978-0-7503-4696-2 (myPrint)
ISBN 978-0-7503-4694-8 (mobi)

DOI 10.1088/978-0-7503-4695-5

Version: 20221201

IOP ebooks

British Library Cataloguing-in-Publication Data: A catalogue record for this book is available from the British Library.

Published by IOP Publishing, wholly owned by The Institute of Physics, London

IOP Publishing, No.2 The Distillery, Glassfields, Avon Street, Bristol, BS2 0GR, UK

US Office: IOP Publishing, Inc., 190 North Independence Mall West, Suite 601, Philadelphia, PA 19106, USA

Contents

Preface **viii**

Author biographies **ix**

1 Introduction **1-1**

1.1 Interface-driven properties using neutron/x-ray reflectometry 1-1
1.2 Emerging phenomena at interfaces using polarized neutron reflectometry 1-6
1.3 Thin-film growth mechanisms 1-11
1.4 Thin-film deposition techniques 1-14
1.4.1 Resistive heating 1-15
1.4.2 Electron beam heating 1-16
1.4.3 Molecular beam epitaxy (MBE) 1-16
1.4.4 Magnetron sputtering 1-17
1.4.5 Pulsed laser beam deposition (PLD) 1-19
1.4.6 High power pulsed magnetron sputtering (HiPPMS) 1-20
1.5 Other complementary surface characterisation techniques 1-20
References 1-22

2 Theory of neutron reflectometry **2-1**

2.1 Introduction 2-1
2.2 Optical theory of reflection 2-3
2.3 Specular NR using unpolarized neutrons 2-6
2.3.1 Fresnel reflectivity of an ideally flat surface 2-8
2.3.2 Reflection from a film of finite thickness and a periodic multilayer 2-11
2.4 Specular polarized neutron reflectometry 2-15
2.4.1 PNR without spin polarization analysis 2-18
2.4.2 PNR with spin polarization analysis 2-21
2.5 Coherence area and resolution in neutron reflectometry experiments 2-26
2.6 Off-specular (or diffuse) neutron scattering 2-30
2.7 Polarized diffuse neutron scattering 2-35
2.7.1 Polarized DNS without spin polarization analysis under the Born approximation 2-35
2.7.2 Polarized DNS with spin polarization analysis under DWBA 2-38
2.8 Data analysis for reflectometry 2-39
References 2-40

3 Understanding emerging phenomena at interfaces using specular neutron and x-ray reflectometry 3-1

3.1 Emerging phenomena at interfaces and characterization 3-1
3.2 Ferromagnetic metal/semiconductor heterostructures 3-5
3.2.1 Asymmetric and reduced interfacial magnetization in the Fe/Ge multilayer 3-6
3.2.2 Different structural and magnetic roughnesses at the interfaces of Fe/Ge heterostructures 3-7
3.2.3 Formation of nanocrystalline alloy phase with large magnetic anisotropy at interfaces on annealing of the Fe/Ge multilayer 3-9
3.3 Interlayer exchange coupling: ferromagnetic metal/nonmagnetic metal heterostructures 3-12
3.3.1 Bilinear exchange coupling 3-12
3.3.2 Biquadratic exchange coupling 3-14
3.4 Multilayer with noncolinear and chiral magnetic structures 3-17
3.4.1 Transition metal/transition metal multilayers 3-21
3.4.2 Magnetic phase transitions in rare-earth/transition metal multilayer 3-22
3.4.3 Rare-earth/rare-earth heterostructures 3-28
3.5 Exchange bias: interface magnetization 3-29
3.5.1 Exchange bias in FM/AFM core–shell systems 3-32
3.5.2 Exchange bias in transition metal (TM)/TM oxide heterostructures 3-32
3.5.3 Exchange bias in other non-oxide-based heterostructures 3-32
3.5.4 Exchange bias in rare-earth/transition metal AFM-coupled heterostructures 3-34
3.5.5 Exchange bias in complex oxide heterostructures 3-35
3.6 Proximity effect and coupling in complex oxide magnetic and superconducting heterostructures 3-35
3.7 Strain-driven interfacial magnetism in complex oxide heterostructures 3-40
3.8 Emergent and interface-induced magnetism at complex oxide interfaces 3-44
3.9 Superdense and nonmagnetic Co phase at interfaces 3-45
3.10 Tracking interdiffusion and self-diffusion kinetics at interfaces 3-47
3.10.1 Ni/Ti multilayer 3-50
3.10.2 Ni/Al multilayer 3-51
3.10.3 FePt/Cu multilayer 3-53
3.10.4 Self-diffusion in isotopic multilayers 3-54

3.11 Proximity-driven magnetic order in topological insulators and interface magnetization in Weyl semimetal 3-55

3.12 Control of local magnetization in isovalent oxide heterostructures by interface engineering 3-55

References 3-56

4 Correlation of interface morphology and magnetism in heterostructures: off-specular (diffuse) scattering 4-1

4.1 Off-specular (diffuse) scattering 4-1

4.2 Off-specular (diffuse) x-ray scattering 4-5

4.2.1 Interface morphology of soft matter heterostructures 4-6

4.2.2 Vertically correlated interface morphology in an organic semiconductor multilayer 4-6

4.2.3 Correlation between surface rumpling, structural phase transformation, and surface morphology in $SrTiO_3$ 4-9

4.2.4 Correlation of charge (structure) and magnetic roughness in Fe/Gd multilayers 4-11

4.2.5 Correlation between charge and magnetic morphology in manganite film 4-11

4.3 Off-specular (diffuse) neutron scattering 4-13

4.3.1 Magnetic domain disorder and its correlation in anti-ferromagnetically coupled multilayers 4-15

4.3.2 Corrosion-driven structure and magnetic roughness in Ni films 4-18

4.3.3 Helical magnetic structure and evolution of magnetic domains in rare-earth-based multilayers 4-24

References 4-27

5 Summary and outlook 5-1

Preface

This book presents x-ray and neutron reflectometry techniques and how they can unravel interface structures and magnetism at the mesoscopic length scale in thin films and multilayers. The aim is to introduce the reader to neutron and x-ray reflectometry in general with an emphasis on polarized neutron reflectometry for magnetic properties that are strongly affected by interface characteristics. The field of ultra-thin films, a subgroup of nanomaterials, has seen a huge upsurge since the 1990s. The Nobel prize for the discovery of giant magnetoresistance stands witness to the importance of tunable magnetic coupling between magnetic layers. At present, a very large number of studies are being carried out towards designing thin film heterostructures and understanding their structure, composition, and magnetism at interfaces, targeting the discovery of novel materials. Emerging properties and magnetism at the interfaces in such heterostructures are of deep research and application interest. This book is an attempt to demonstrate the important role of neutron (and x-ray) reflectivity measurements in understanding such properties.

The book is divided into two main parts. The first part consists of chapters 1 and 2, which introduce the techniques of x-ray and neutron reflectometry for studies related to thin film heterostructures and briefly mention various studies that are being carried out using these techniques. This first chapter also introduces the reader to various thin film deposition techniques because they are part and parcel of the studies being carried out. The second chapter presents the theory of specular and off-specular reflectometry along with a brief mention of model-based data analysis techniques.

The second and central theme of this book is the elaboration of the importance of techniques, with a large number of examples taken from recent work on materials of basic and technological interest, where the interface properties have been resolved at a sub-nanometre length scales. Chapters 3 and 4 discuss the specular and off-specular scattering, respectively, with a large number of references to studies done by the authors and other researchers.

The topics and the results presented are attempts to describe the emerging physical and magnetic properties at the interface of two materials of different natures, indicating that novel properties often emerge at the interfaces that are not present in the adjacent layers. Interface coupling affects the physical and magnetic structure in the layers comprising a heterostructure. During the 1990s the technique of neutron reflectometry (and x-ray reflectometry) emerged as one of the primary tools to understand structure and magnetism in thin film heterostructures.

One of the authors (SB) is grateful to Professor Sunil Sinha for introducing him to the subject of neutron reflectometry during a summer school in 1990. But for his gentle nudge, the author would not have dived into the field and gathered such rich bounty.

Author biographies

Saibal Basu

Saibal Basu has carried out research in the field of neutron scattering and neutron instrumentation at the Dhruva reactor in India for more than three decades. He was primarily responsible for developing a polarized neutron reflectometer at the Dhruva reactor. He headed the neutron scattering group at Bhabha Atomic Research Centre and was a professor at Homi Bhabha National Institute (deemed to be a university). Based on his experience in neutron research, he worked as a research reactor specialist at the International Atomic Energy Agency in 2001. He has carried out extensive neutron reflectometry research at several major neutron scattering centres. These include the Dhruva reactor in India, the ISIS Spallation Neutron Source, the Rutherford Appleton Laboratory in the UK, and the NIST Centre for Neutron Research (NCNR) at NIST, MD, USA. His work mainly targets interface properties such as magnetism and interface alloy formation in thin film heterostructures.

Surendra Singh

Surendra Singh is a scientist at Solid State Physics Division, Bhabha Atomic Research Centre, Mumbai. He is also a faculty at Homi Bhabha National Institute (Deemed to be University) Mumbai. He received his PhD in physics from the University of Mumbai in the year 2008. He was a post-doctoral fellow at Los Alamos National Laboratory, USA during 2010–12. He has been instrumental in setting up the polarized neutron reflectometer at Dhruva reactor, BARC for studying magnetic properties of heterostructures and superlattices. His current research interests is in the field of structure and magnetic properties of complex oxide interfaces and superlattices. He has authored/co-authored over 100 research publications in high-impact international journals.

IOP Publishing

Neutron and X-ray Reflectometry
Emerging phenomena at heterostructure interfaces
Saibal Basu and Surendra Singh

Chapter 1

Introduction

1.1 Interface-driven properties using neutron/x-ray reflectometry

Thin-film research has exploded over the past three decades in several directions. In particular, the hope that desirable properties of a material can be change and tailored at lower dimensions has fuelled the search for new materials in thin-film form. Thin films are exploited in many applications [1], ranging from magnetic and semiconductor devices to optical coatings and even in pharmaceuticals (polymers). Thin-film heterostructures are also important for the study of materials with new and unique properties due to the possibility of tuning their crystallographic and morphological properties, particularly at interfaces. By the term 'heterostructure' we refer to thin layers of material or materials of single layers, bilayers, or multiple layers (multilayers) deposited on a chosen substrate. The 2007 Nobel Prize awarded to Albert Fert and Peter Grünberg is a high point in this research [2]. This research has given birth to a new kind of memory chip based on the giant magnetoresistance (GMR) phenomenon [3]. Spintronics is another phenomenon based on the injection of spins across an interface between ferromagnetic (FM) metal/semiconductors etc, which is also being studied extensively in thin-film heterostructures at the moment [4]. Significant advances in thin-film growth techniques have made it possible to grow high-quality heterostructures with precise control of both the structure and chemical composition. Interfaces of high-quality heterostructures are epicentres for many novel properties due to their sharp electronic density modulations, symmetry lowering, and strain as a result of lattice mismatch and defects at the interfaces. In addition, the tunable structural and magnetic properties of thin-film heterostructures along with epitaxial compatibility have provided systems for functional devices such as spin valves, magnetic read heads, and tunnel junctions. This has led to intense research activity into layered heterostructures, specifically to study the role of the interfaces which might lead to novel emergent phenomena.

In the quest for new material properties across an interface in a thin film, a large number of attempts are being made to artificially engineer epitaxial heterostructures,

doi:10.1088/978-0-7503-4695-5ch1

which enable new functionalities that cannot be realized with the individual constituents. Such functionality is derived from exchange interactions across interfaces [5]. To cite a few examples, experimentally a conducting layer at the interface of two oxide insulators has been observed [6], magnetic moment can be transferred at the interface from a ferromagnet to a nonmagnetic material [7, 8], the coexistence of superconductivity and ferromagnetism in two-dimensional nonmagnetic and insulating oxides has been found [9], magnetic dead-layers at the interface of magnetic and superconducting layers have been reported [10, 11] as has the tunnelling of Cooper pairs across an interface from a superconductor to a ferromagnet through an insulator [12, 13] and the phenomena of exchange coupling at the interface of a ferromagnet and an antiferromagnet have been studied widely [14]. The list is certainly not exhaustive but it is an indicator of the importance of the interfaces in dictating properties in thin-film heterostructures. Over the past two decades researchers have observed many properties in thin-film multilayer heterostructures which are driven by the proximity of two components at an interface and the interactions that are not there in a single bulk component. In addition, the widespread application of films with thicknesses from atomic monolayers such as graphene [15] to microns is due to the development of thin-film deposition techniques. The juxtaposition of materials with different electronic, magnetic, and structural properties is key to observing new materials with properties to fuel new technology. In this book we attempt to cover these observations under the broad canopy of 'emerging phenomena' and demonstrate the role of neutron and x-ray reflectometry in correlating these phenomena to the corresponding physical and often magnetic properties at the interfaces. The physical nature of the interfaces has a large bearing on the emerging phenomena that one attempts to monitor and tailor to the desired target property. A brief discussion about the growth of thin-film heterostructures is also given as a part of the introduction.

The neutron, a nuclear particle, was introduced to science in a brief letter to the journal *Nature* by Chadwick in 1932 [16]. In an equally brief paper, Fermi and Zinn demonstrated in 1946 [17] that slow neutron reflection from a surface of a material can be successfully used for the determination of its coherent scattering length for thermal neutrons. Although the stage was set, the technique of 'neutron reflectometry' came of age somewhat later, around the mid-1980s, thanks to the contribution of several researchers, notably Gian Felcher from Argonne National Laboratory [18]. He established possibly the first neutron reflectometer at Argonne National Laboratory to study thin films. The popularity of the technique grew rapidly for studies related to films and interfaces. At present every major neutron source has not one but several variants of instruments dedicated to reflectometry technique for thin-film research studies. Interestingly, Röntgen, the discoverer of x-rays, failed to observe the refraction of the rays by any medium in his investigations. The use of x-ray reflectometry is due to L G Parratt in 1954 [19] when he used the technique to study copper-coated glass. Notably, he also developed the numerical formalism for reflectometry data analysis in a seminal article [19], which is used by all, even today. Currently both of these complementary techniques (neutron and x-ray reflectometry) are popular for studying thin-film structures worldwide.

Neutron reflectometry (NR) follows closely the technique of x-ray reflectometry (XRR) as it is known popularly. Both techniques depend on the reflection of neutrons or x-rays from a surface. The reflection phenomena indicate the wave properties of the incident radiation, the de Broglie wavelength for a thermal neutron in the present case. The technique of neutron (x-ray) reflectometry is fairly simple conceptually and similar to the optical reflection that one encounters in daily life. It is known that the reflection of a wave occurs whenever there is a refractive index change at an interface. We experience routinely our reflection in glass mirrors or a landscape reflected in the waters of a lake. Due to the smooth surface of the mirror and water, the image produced is clear and sharp, whereas the image produced from reflecting surfaces with roughness or ripples is of poor quality. The process of x-ray and neutron reflectometry is an extension of this observation to the mesoscopic or nanometre length scales.

The NR (and XRR) technique, in simple terms, consists of preparing a highly collimated beam of radiation (neutrons/x-rays) and observing the reflection of the beam from the surface as a function of the incident angle at grazing incidence (angle θ_i in figure 1.1(a)). There are two types of reflection occurring from a surface, one that follows Snell's law of reflection, i.e. the angle of incidence on the surface is equal to the

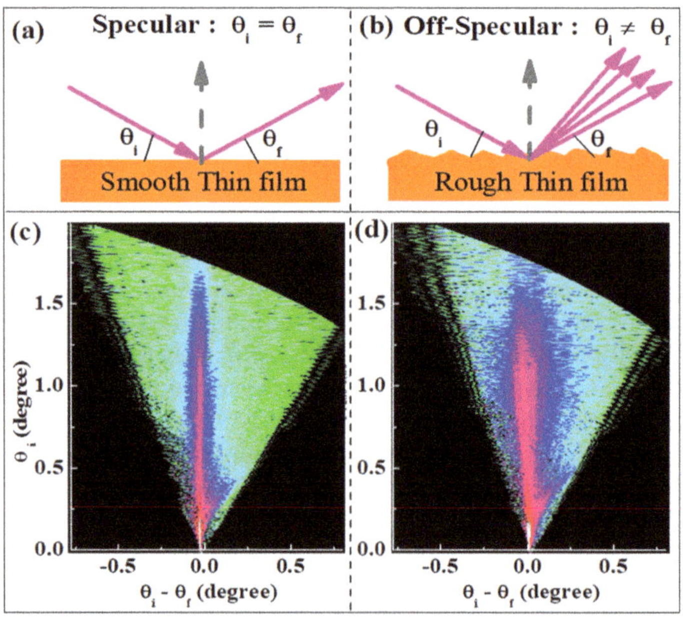

Figure 1.1. Schematic of reflection of neutrons/x-rays with a grazing angle of incidence (θ_i) and angle of reflection (θ_f) from (a) smooth and (b) rough thin film. Generally a specular ($\theta_i = \theta_f$) and off-specular ($\theta_i \neq \theta_f$) reflectivity is observed from smooth and rough surfaces. X-ray reflection from $SrTiO_3$ (STO) single-crystal substrate at a temperature (T): (c) $T = 300$ K and (d) $T = 50$ K. STO exhibits a phase transition at a $T = 105$ K and the surface shows rumpling with a rough interface. We have specular reflectivity along the y-axis at $\theta_i - \theta_f = 0$. A smooth surface ($T = 300$ K) shows sharp reflectivity, whereas the rough surface ($T = 50$ K) shows a spread of the beam around $\theta_i - \theta_f = 0$ with low intensity.

angle of reflection ($\theta_i = \theta_f$, as shown in figure 1.1(a)), which is known as specular reflection, and the other when the equality of incident and reflection angles is violated ($\theta_i \neq \theta_f$, in figure 1.1(b)), known as off-specular (diffuse) reflection. Both reflections are important to probe thin films and will be discussed in due course. Figures 1.1(a) and (b) show schematics for specular and off-specular reflection conditions, respectively. X-ray reflection from a smooth (high temperature, 300 K) and rough (low temperature, 50 K) $SrTiO_3$ (STO) substrate [20] is depicted in figures 1.1(c) and (d), respectively. In brief, observing the reflected neutron and x-ray beams helps one to unravel various structural parameters of thin films and multilayers, which is the topic of the current book.

The immense popularity of the reflectometry technique comes from the fact that it is a non-destructive technique with an excellent spatial depth resolution in the sub-nanometre length scale for probing physical properties in thin-film heterostructures. The schematic of a heterostructure of two materials *A* and *B* grown on a substrate is shown in figure 1.2(a). Figure 1.2(b) shows the high-resolution cross-sectional transmission electron microscopy (TEM) image of the heterostructure of two such materials. In a heterostructure, the interface of the two layers (figure 1.2) is of utmost importance in dictating the properties of the heterostructure. The reflectometry technique is ideally suited for probing such nanostructures and multilayers. We will demonstrate that x-ray and unpolarized neutron reflectometry can provide a density profile as a function of depth in ultra-thin nanolayers with sub-nanometre resolution, together with interface composition and roughness. The interface between the components of two layers in a thin film is never smooth or two-dimensional at an atomic length scale, in reality. The interface has an associated interface roughness. The roughness may be considered a fluctuation in the mean height across the plane of the interface that is visible in an atomic force microscopy image at a film–air interface. Figure 1.3 shows an atomic force microscopy image of size 5 μm × 5 μm of a metal film. The graph in figure 1.3 shows the histogram (height distribution) of the atomic force microscopy image and the FWHM of the Gaussian fit (solid line) to the histogram data (open triangles) is termed the

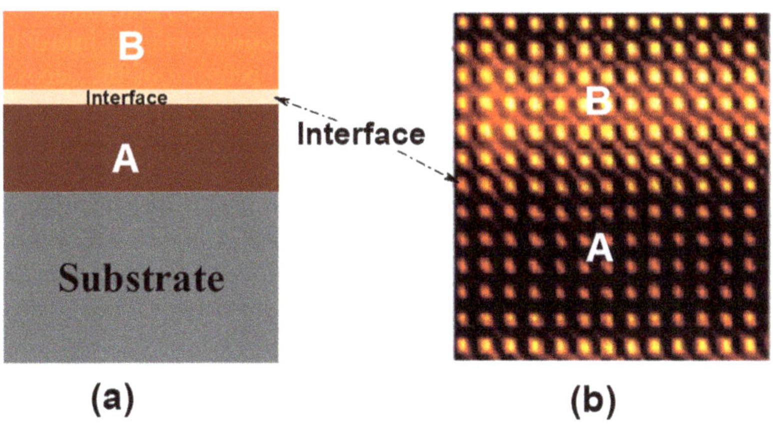

Figure 1.2. (a) Schematic of the *A*/*B* heterostructure, where a thin film of material *A* is first grown on the substrate and then material *B* is grown on material *A*. The *A*/*B* (*B* on *A*) interface is also shown. (b) High-resolution TEM cross-section image of a heterostructure.

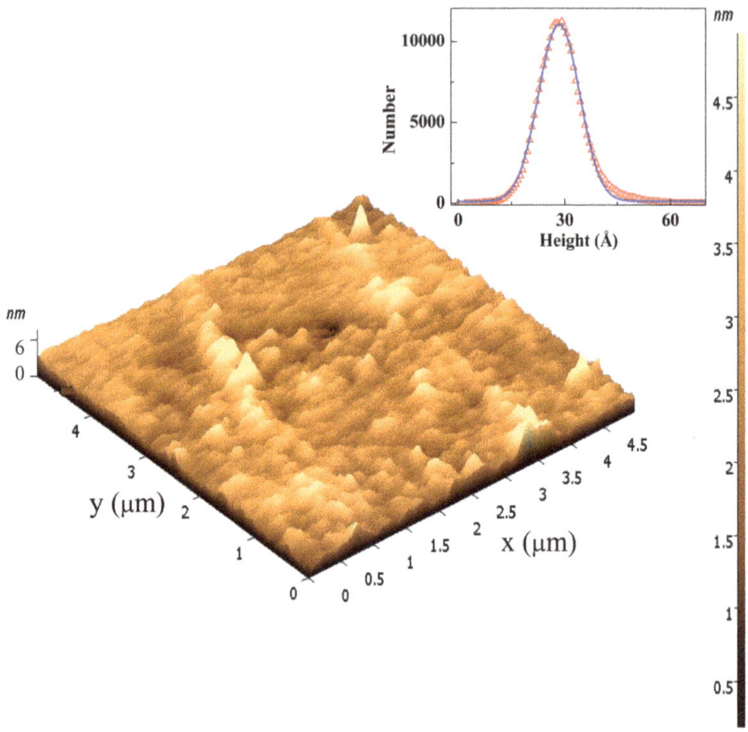

Figure 1.3. Atomic force microscopy image of a film grown by sputtering. The graph shows the height distribution (open triangles) of the image of the film–air interface (surface) and the FWHM of the Gaussian fit (solid line) provides the interface roughness.

interface roughness in Å. The roughness depends strongly on the surface energy of the components on either side of the interface. If we call the two components (materials) *A* and *B*, then while a heterostructure is deposited, the interface for *A* deposited on *B* and an interface for *B* deposited on *A* will have different roughness and surface topologies due to the variation in surface energy. Moreover, there will be interdiffusion of the components during deposition, based on the temperature of the deposition and often on the temperature of annealing after deposition. The kinetic energy of the component atoms/molecules impinging on the substrate, or on layers already deposited, varies from one deposition technique to another. There will also be defects at the interface that depend on the process of deposition and the components. The emerging phenomena at the interfaces are a convolution of all the above physical processes. The present book attempts to demonstrate the role of reflectometry techniques to identify the phenomena and their correlation with the above parameters in various studies.

In addition, polarized neutron reflectometry (PNR) is capable of tracking magnetization depth profiles with a resolution in nanometres. PNR with polarization analysis for the reflected beam also resolves planar magnetic structure at a mesoscopic length scale. Because of the excellent spatial resolution, one can obtain information about hidden layers and interfaces from reflectometry experiments, which is otherwise not possible by any other surface-sensitive technique. The closest one can consider in terms of resolution is transmission electron microscopy (TEM), which is a destructive technique. Unlike macroscopic magnetization techniques (e.g. the superconducting quantum interference device (SQUID), vibrating sample magnetometer (VSM), etc), which are only sensitive to the average magnetization of the whole sample (including the substrate—the contribution from the substrates is a serious concern in interfaces and heterostructures), PNR, using its intrinsic properties, can probe the magnetic moment density of buried interfaces with sub-nanometre depth resolution in a non-invasive manner.

1.2 Emerging phenomena at interfaces using polarized neutron reflectometry

The following are a few emerging phenomena at interfaces, where the PNR technique has played an important role in the investigation and understanding of these phenomena.

A. Investigation of interface magnetization in ferromagnet/semiconductor heterostructures
Ferromagnetic metal–semiconductor interfaces exhibit interesting physical and chemical properties. After Datta and Das [21] proposed an electron wave analogue of the electro-optic light modulator, a new generation of electronic devices, named spintronics, exploiting the spin degree of freedom of carriers in ferromagnet/semiconductor (FM/S) interfaces [4] were envisaged for a range of applications, namely spin memories, spin field-effect transistors, spin transistors, etc [22, 23]. FM/S heterostructures, e.g. 3*d* transition metal FMs (Fe, Co and Ni, etc) on a semiconductor (Si, Ge and GaAs, etc), are promising systems for the transfer of magnetic or spin-related wavefunctions into semiconductor materials, a necessity for the realization of spintronic devices. Extensive work has been carried out to effectively generate and detect spin accumulations in semiconductors [24, 25] and metals [26, 27] for the last 20 years. However, the property of interest in the possible spintronics systems (FM/S interfaces), i.e. the transport of spins across the FM/S interface, depends strongly on interface properties such as roughness, interdiffusion, alloying, and interface magnetization. In addition, nonmagnetic alloy formation at the interfaces of the FM/S heterostructures has been detrimental to the electron spin injection from the FM to the semiconductor. Therefore, investigation of interface magnetization in FM/S heterostructures is key for the realization of spintronic devices. The Fe/Ge heterostructure is one of the most extensively studied systems, using a variety of experimental and theoretical techniques [28–31], where spin–charge interconversion at the interface between the two materials was observed due to the presence of Rashba states [32], which controlled spin accumulation. Using the PNR technique in combination with XRR and macroscopic magnetization measurements, Fe/Ge heterostructures have been studied [33–36] with a special emphasis on the investigation of the interface magnetization and understanding the role of structure and magnetic roughness on the depth-dependent magnetic properties. PNR results for Fe/Ge heterostructures showed a reduction in magnetization for the interfacial Fe layer because of the interdiffusion of atoms at the interfaces.

B. Investigation of alloy formation at interfaces due to interdiffusion in ultra-thin films and isotopic diffusion

On annealing metal/metal heterostructures the interdiffusion of atoms at interfaces take place, which provides an excellent route to produce intermetallic alloy phases at interfaces. In the case of ultra-thin films, such alloying at the interface occurs at much lower temperatures compared to the melting temperatures of the constituent elements. Ni- and Fe-based multilayers (e.g. Ni/Ti, Ni/Al, Ni/Ge, FePt/Cu, etc) have been studied extensively for their structural and magnetic properties. The studies on intermetallic alloys of these systems have largely been focused on their physical and mechanical properties, particularly involving low-temperature ductility and high-temperature strength-based technological applications [37]. XRR and PNR techniques were used to investigate interdiffusion and the composition of alloy formation at the interfaces of Ni/Ti [38], Ni/Al [39–42], Ni/Ge [43], and FePt/Cu [44–46] heterostructures on annealing of the heterostructures at different temperatures. These techniques were also used to study the structure and magnetic properties of these intermetallic alloys. The PNR results suggested the formation of a non-magnetic intermetallic alloy phase at the interface between the Ni and Ti, Ni and Al, and Ni and Ge interfaces, whereas a ferromagnetic alloy phase at the interfaces was observed in the case of FePt/Cu heterostructures.

C. High-density and nonmagnetic Co film at interfaces

Studies of the structure and magnetism of 3*d* transition metals (Fe, Ni, and Co) have remained interesting beyond the traditional bulk materials and thin films, particularly at reduced volume (under high pressure), and have received a great deal of attention [47]. Theoretically, the magnetic moment of Co decreases with decreasing atomic volume and the magnetic moment of a face-centred-cubic (fcc) Co phase nearly vanishes when the atomic volume is reduced by a factor of ~1.4 [48]. However, there was no experimental study in the literature on the formation of high-density Co film showing nonmagnetic properties, although reports have suggested the formation of a film with compressive stress due to polycrystalline growth, which can increase the density of the film [49]. A high-density nonmagnetic fcc Co phase at interfaces was investigated primarily by PNR and XRR [50]. The experimental evidence of non-magnetic superdense Co at the interfaces of Co film on a Si substrate indicates the possibility of the existence of nonmagnetic Co in the Earth's core under high pressure.

D. Investigation of helical magnetic structure at interfaces between a rare earth and transition metal

The compensated rare earth (RE)–transition metal (TM) alloys and heterostructures are considered potential candidate materials for realizing memory devices with higher speed and density [51, 52]. In addition, nanosized stable magnetic helical structures in RE/TM multilayers were theoretically predicted to be used for magnetic energy storage [53, 54]. Magnetization reversal and a magnetic helical configuration in RE–TM multilayers with no external magnetic field around the compensation temperature (T_{comp}, the temperature at which the total moments of the RE–TM multilayer tend to zero due to the strong antiferromagnetic (AF) interaction at interfaces) is the key to manipulating magnetic devices [55, 56]. PNR measurements from the Gd/Co multilayer as a function of temperature and field demonstrated the formation of a helical magnetic structure in the form of planar 2π domain walls (DWs) within both the layers of Gd and Co near T_{comp} [57–59]. The detailed magnetic structure obtained from PNR further confirmed an AF coupling between the Gd and Co layers and suggested that AF coupling and the formation of planar 2π DW below T_{comp} are responsible for the observed negative exchange bias.

E. Magnetization at the interface between two nonmagnetic and insulating oxides
Artificial heterostructures of transition metal oxides are ideal systems to explore interfacial effects that could lead to new phases. Ohtomo and Hwang [6] reported unusually high conductivity at the interface of nonmagnetic and insulating $LaAlO_3$ (LAO) and $SrTiO_3$ (STO) heterostructures. The existence of long-range magnetism at the LAO/STO interface was first inferred from the magnetoresistance (MR) measurements of the LAO/STO interface at low temperatures and high magnetic fields [7, 9]. To investigate the interface magnetization of LAO/STO heterostructures, PNR measurements were carried out on several LAO/STO superlattices at a very low temperature (~1.7 K) and a high magnetic field of ~11 T [8]. However, the PNR results suggested the highest magnetization of 2G for LAO/STO interfaces [8].

F. Ferromagnetism at the interface between an antiferromagnetic insulator and paramagnetic metal oxides
Interface-driven novel phenomena in heterostructures, e.g. ferromagnetism at the interface of two non-FM materials, are of interest both from a fundamental perspective and also for potential applications associated with spintronics. A ferromagnetic spin order was observed for $LaCrO_3$(LCO)/$LaFeO_3$(LFO) superlattices using macroscopic magnetization measurements, where both the LCO and LFO oxides are AF insulators [60]. Other AF insulators systems are $LaMnO_3$ (LMO) and $SrMnO_3$ (SMO) where the $(LMO)_{2n}/(SMO)_n$ superlattices showed FM metallic character for a short superlattice period ($n < 3$) via double exchange interaction [61]. Using the depth sensitivity of PNR measurements an interfacial FM order was observed in $CaRuO_3$ (CRO)/$CaMnO_3$ (CMO) superlattices, where CRO and CMO are a paramagnetic metal and an AF insulator, respectively [62]. The system exhibited the emergence of exchange bias in this system as a consequence of the exchange coupling of interface-induced FM and the rest of the antiferromagnetic CMO layer. PNR also confirms the existence of FM at the interface of $LaNiO_3$ (LNO)/CMO superlattices [63], where LNO is a paramagnetic metal.

G. Ferromagnetism at the interface between ferromagnet and antiferromagnet multiferroic oxides
Several fascinating interfacial phenomena have been observed in heterostructures involving FM and AF oxides such as $La_{0.7}Sr_{0.3}MnO_3$ (LSMO) and $BiFeO_3$ (BFO), respectively. LSMO is a spin-polarized manganite with a magnetic transition (Curie temperature, T_c) of ~350 K, while BFO is a single-phase multiferroic material that exhibits magnetoelectric coupling between ferroelectric (FE) (FE transition temperature

$T_c \sim 1103$ K) and AF ($T_N \sim 643$ K) order parameters [64]. A weak FM for BFO (~0.03 μ_B) as a result of canting of the AF structure due to the Dzyaloshinskii–Moriya interaction has been suggested theoretically [65]. An enhanced magnetic moment in the interfacial BFO layer in the BFO/LSMO heterostructure was observed, which was attributed to Fe–Mn hybridization and orbital reconstruction [66]. Using PNR, an induced magnetization of 75 ± 25 kA m^{-1} for the interfacial BFO layer extending from the interface through several atomic layers of the BFO at 10 K in an LSMO/BFO superlattice was observed [67]. The induced magnetization in BFO was explained by density functional theory suggesting that the size of the bandgap of BFO plays an important role. A PNR study also suggested different magnetic couplings [68], ferromagnetic [67, 69] and antiferromagnetic [70], for two BFO/LSMO heterostructures, which might have resulted due to different surface terminations of LSMO in these heterostructures.

H. Investigation of a magnetic depleted layer at the interface between ferromagnetic and superconducting complex oxides

The heterointerfaces of oxide superconductors (SCs) and FM lead to charge-transfer-driven orbital ordering and reconstructions. The orbital reconstruction was first observed at the $YBa_2Cu_3O_7$ (YBCO)/$La_{2/3}Ca_{1/3}MnO_3$ (LCMO) interface [71]. LCMO is a half-metallic FM and YBCO is a d-wave SC. LCMO/YBCO heterostructures have been extensively studied as a model system to investigate the phase competition between the FM and SC states. The studies suggested the magnetic proximity effect (MPE) between SC and FM in these oxide films. PNR measurements suggested a reduction and modulation in magnetization at the interfaces of the LCMO/YBCO heterostructure below the superconducting transition temperature (T_{SC}) [10, 72] of YBCO. A suppression of magnetism in the whole LCMO layer of thickness ~8.5 nm was observed below T_{SC} [10]. PNR results also indicated the formation of a magnetic depleted (or dead) (MD) layer of a thickness of ~1 nm to 6 nm in LCMO/YBCO [11] and LSMO/YBCO [13] heterostructures below T_{SC}. The results are the signature of long-range MPE in these cuprate–manganite oxide heterostructures.

I. Investigation of the strength of the magnetic proximity effect through the tunnelling of Cooper pairs across superconductor/insulator/ferromagnet oxides

The magnetic proximity effect was first investigated by PNR measurements in the so-called tunnelling geometry of an FM/insulator (I)/SC heterostructure, where the FM and SC were separated by an insulator [12]. PNR results suggested the emergence of an MD layer at the interface of a YBCO/STO/LCMO heterostructure below T_{SC} [12], which was attributed to tunnelling of Cooper pairs from the SC to the FM across an insulator (STO), that modifies the interfacial magnetism in LCMO. The study also showed that on decreasing the thickness of the insulator layer in the FM/I/SC heterostructure, the thickness of the MD layer at the LCMO/STO interface increases. These results were later confirmed by Paull *et al* [73], where PNR results suggested the disappearance of the MD layer on increasing the applied magnetic field below T_{SC} (the large field destroyed the SC state and stopped the tunnelling of Cooper pairs responsible for interfacial magnetic modulation). Using spin-dependent PNR experiments an unusual proximity behaviour with an unexpected magnetic reversal confined to the interface region of the LSMO layer below the T_{SC} was demonstrated in YBCO/STO/LSMO heterostructures [13]. A recent study suggested that the proximity effect was suppressed by irradiating the trilayer YBCO/STO/LSMO [74]. Modification in magnetization of the interfacial LSMO layer measured by PNR below the T_{SC} in these trilayers clearly suggested the tunnelling of Cooper pairs from the SC to FM oxides through an insulator.

J. Modified magnetic structure due to interlayer coupling

The interlayer coupling between magnetic/nonmagnetic heterostructures plays an important role in governing their magnetic properties. The coupling may initiate from the exchange and magnetostatic interactions affecting magnetizations of ferromagnetic films across a nonmagnetic spacer layer. In the case of a metallic spacer layer, the interlayer exchange coupling (IEC) oscillates as a function of its thickness, while in the case of an insulating spacer layer the IEC decays exponentially, reflecting the evanescent nature of the exchange-mediated states. PNR experiments helped to investigate the AF arrangement of the Fe layers in Fe/Cr multilayers way back in 1991 [75]. Later two significant studies with PNR measurements on interlayer coupling were reported: (i) the identification of oscillatory exchange coupling in Co/Cu superlattices [76] and (ii) noncollinear and collinear magnetic structures in exchange-coupled Fe/Cr superlattices [77]. Further quantification of magnetic domain disorder and its correlations in antiferromagnetically coupled multilayers was studied using both specular and off-specular PNR [78, 79].

K. Exchange bias in ferromagnet/antiferromagnet and ferromagnet/non-magnet heterostructures

The exchange bias (EB) is due to an interaction between an FM and an AF layer [14] and it is manifested as a shift and a broadening of the hysteresis loop in magnetization measurements from the FM/AF heterostructures. Ball *et al* [80] used PNR in 1995 to investigate the depth-dependent magnetization profiles of the exchange bias system. Thereafter many studies on FM/AF heterostructures using PNR were reported for understanding exchange interaction-driven magnetic properties such as the layer-by-layer magnetization of a double-superlattice [81], asymmetric magnetization reversal [82], training effect [83], and domain formation due to magnetic reversal [84]. There are many studies on exchange bias phenomena in FM/AF heterostructures using PNR. Recently there have been many studies where an FM/non-magnet heterostructure, in particular strongly correlated electron oxide systems, has shown EB as a result of interface-dependent phenomena. In this context, interface engineering in complex oxide heterostructures has provided access to hidden phases as well as induced complex magnetic structures in a nonmagnetic material. EB was observed in many oxide heterostructures, e.g. paramagnetic LNO–FM LMO superlattices [85], paramagnetic metal CRO–AF insulator CMO heterostructures [62], LNO–CMO heterostructures [63], LNO–LSMO heterostructures, etc [86], where PNR has played an important role to reveal the magnetic phase formed at the interfaces.

L. Reverse magnetization at the interface of an electrodeposited ferromagnetic film

The PNR technique was successfully used to reveal the magnetic properties of two Co films grown by different techniques, namely sputtering and electrodeposition

(ED), which provide different nucleation and growth mechanism for the deposition of thin films [87]. PNR results revealed a uniform magnetization of sputtered Co film, while the Co film grown by ED suggested a two-layer structure for Co film. The film–air interface region of the Co film showed low-density Co with magnetization opposite to the rest of Co film for the ED-grown sample. Thus PNR results from ED-grown Co film suggested an AF coupling between the surface and bulk Co phase which was also responsible for anomalous magnetic transport behaviour.

M. Ferromagnetic topological insulating phase formation at the interface due to proximity coupling

Topological insulators (TIs) are insulating materials that display conducting surface states, which opens up many technological application possibilities. Inducing ferromagnetic order in TIs using the magnetic proximity effect can produce new physical phenomena that may have technological applications. Using spin-dependent PNR measurements, Katmis *et al* [88] observed ferromagnetism for interfacial (thickness ~2 nm) Bi_2Se_3 TI in a EuS (FM)/Bi_2Se_3 (TI) heterostructure. The coupling of the FM EuS and the TI Bi_2Se_3 also suggested ferromagnetism for the system at room temperature, a much higher temperature than the T_c of the bulk EuS.

These emerging phenomena clearly suggested that the PNR techniques have been successfully used to investigate new physical phenomena at the interfaces of heterostructures that have technologically important consequences. However, growing high-quality heterostructures is also a very important part of interface engineering and a brief introduction to the growth mechanisms and techniques is provided in the following sections. The growth of reflectometry techniques for probing thin films at major neutrons has happened in parallel with the growth of thin-film research and the development of numerous thin-film growth techniques.

1.3 Thin-film growth mechanisms

At this point, it is relevant that we briefly present various thin-film growth mechanisms that are important for interface engineering. We will discuss the mechanisms when more than an atomic monolayer grows on a substrate of choice. The physical process of thin-film growth depends on interactions between the atoms being deposited (adatoms) and the interaction between the adatoms and the substrate surface. Depending on these interactions, the growth of a film can be broadly classified into three categories:

i. *Frank Van der Merve (FVM) mode or layer-by-layer growth* (figure 1.4(a)): This occurs when the bonding of the atoms of the film with the substrate is stronger compared to interatomic bonding. In this case, it is energetically favourable for the film atoms to cover the substrate layer by layer due to better wettability.
ii. *Vollmer–Weber (VW) or island growth* (figure 1.4(b)): This is the opposite of the FVM mode when the interatomic bonding is stronger than atom–substrate bonding. In this case, the film grows in islands of several atomic layers on the substrate.
iii. *Stranski–Krasnatov (SK) mode* (figure 1.4(c)): This is intermediate between the FVM and VW modes. In this mode after a few atomic layers have grown, the islands start growing on these layers.

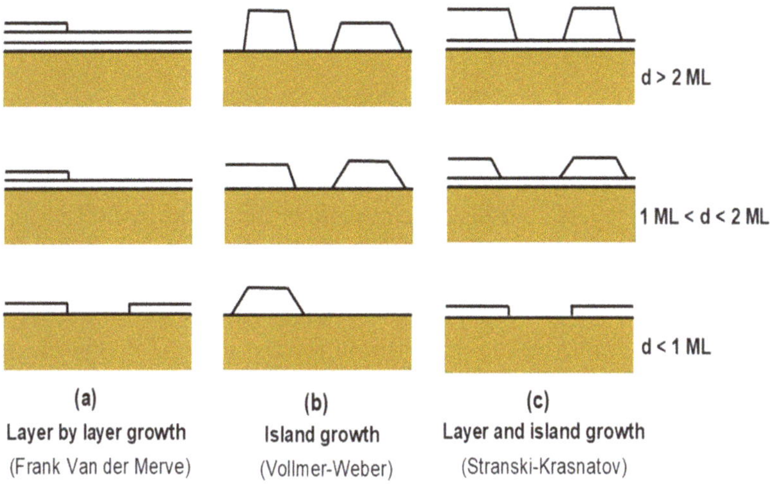

Figure 1.4. Classical growth modes of thin films.

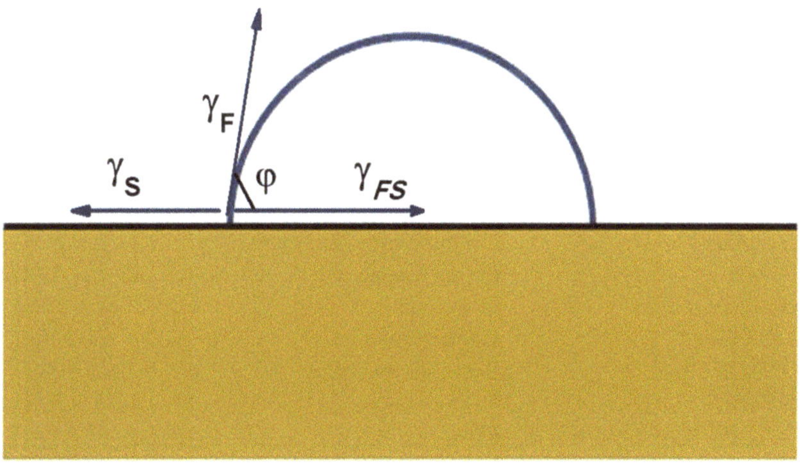

Figure 1.5. Representation of thin-film nucleation.

The growth modes can be understood in terms of the surface energy or surface tension of the surface and the adatom islands. If γ_S is the surface energy for the substrate, γ_F that for the film, and γ_{FS} the interfacial energy for the film and the substrate, then from figure 1.5, where φ is the contact angle between a film island and the substrate, one can write the energy balance:

$$\gamma_S = \gamma_F \cos\varphi + \gamma_{FS}. \tag{1.1}$$

When the film wets the surface completely, then the contact angle φ is zero and the condition for layer-by-layer growth or the FVM mode is

$$\gamma_S \geqslant \gamma_F + \gamma_{FS}. \tag{1.2}$$

On the other hand, for island or VW growth, the contact angle φ is greater than zero and the energy condition becomes

$$\gamma_S \leqslant \gamma_F + \gamma_{FS}. \tag{1.3}$$

For the intermediate case of the SK mode the energetics change after the intermediate layer/layers have grown and the above relations might switch for the growth of islands on layers. One needs to remember that apart from the energetics of the film growth, the kinetics of film deposition, depending on the technique/techniques used, may cause a metastable film composition/structure at the interface. Closely related to the energetics, another important issue is the atomic structure of the substrate surface. The choice of substrate for the growth of thin films is quite wide and varied today. Often glass or single-crystal Si wafers are used for the deposition of films. Many other single-crystal substrates are available commercially and are being used routinely for growing epitaxial films. A list of such commercially available single-crystal substrates with lattice parameters is given in table 1.1. Regarding the registration of the adatoms on the surface of the substrate, it is

Table 1.1. Commercially available single-crystal substrates used for the growth of epitaxial thin films.

Single-crystal substrates	Crystal structure	Lattice constant (Å)
Si	fcc (Diamond)	$a = b = c = 5.43$
SiO_2	Trigonal	$a = 5.02$, $c = 5.52$
Al_2O_3	Hexagonal	$a = b = 4.75$, $c = 12.99$
$LuAlO_3$ (LuAO)	Orthorhombic perovskite	$a = 5.12$, $b = 5.36$, $c = 7.36$
$YAlO_3$ (YAO)	Orthorhombic	$a = 5.22$, $b = 5.38$, $c = 7.45$
$LaSrAlO_4$ (LSAO)	Hexagonal	$a = 3.75$, $c = 12.63$
$NdAlO_3$ (NAO)	Cubic perovskite	$a = b = c = 3.76$
$LaAlO_3$ (LAO)	Cubic perovskite	$a = b = c = 3.81$
$NdGaO_3$ (NGO)	Orthorhombic	$a = 5.47$, $b = 5.61$, $c = 7.80$
$(LaAlO_3)_{0.29}$–$(SrAl_{1/2}Ta_{1/2}O_3)_{0.71}$ (LSAT)	Cubic	$a = b = c = 3.87$
$LaGaO_3$ (LGO)	Orthorhombic	$a = 5.52$, $b = 5.49$, $c = 7.77$
$SrTiO_3$ (STO)	Cubic perovskite	$a = b = c = 3.90$
$DyScO_3$ (DSO)	Orthorhombic	$a = 5.44$, $b = 5.71$, $c = 7.89$
$TbScO_3$ (TSO)	Orthorhombic	$a = 5.45$, $b = 5.72$, $c = 7.91$
$GdScO_3$ (GSO)	Orthorhombic	$a = 5.45$, $b = 5.75$, $c = 7.93$
$EuScO_3$ (ESO)	Orthorhombic	$a = 5.60$, $b = 5.81$, $c = 8.02$
$SmScO_3$ (SSO)	Orthorhombic	$a = 5.55$, $b = 5.81$, $c = 8.04$
$KTaO_3$ (KTO)	Cubic	$a = b = c = 4.03$
$NdScO_3$ (NSO)	Orthorhombic	$a = 5.60$, $b = 5.84$, $c = 8.08$
$BaTiO_3$ (BTO)	Tetragonal	$a = b = 4.00$, $c = 4.20$
$PrScO_3$ (PSO)	Orthorhombic	$a = 5.64$, $b = 5.85$, $c = 8.11$
$LaLuO_3$ (LLO)	Orthorhombic	$a = 5.82$, $b = 6.01$, $c = 8.38$
MgO	Cubic	$a = b = c = 4.21$

important whether the surface is crystalline or amorphous. The oriented growth of a film on a single-crystal substrate is categorized as epitaxial growth.

If the film material and the substrate belong to the same crystalline structure it is known as a homoepitaxy. When they have a different structures it is known as a heteroepitaxy. The registry of the atoms on the substrate is a function of kinetics as well as the thermodynamics of the growth. Epitaxial films grow with the crystalline registry on a crystalline substrate. If the lattice parameters between the substrate and film are marginally different then the film may grow with a strain which will alter the lattice parameters and properties such as magnetism, band structure, etc. Beyond a certain thickness, the strained film might relax, giving rise to defects at the interfaces. The importance of the deposition mechanism will be evident in our later discussion on interface properties and their dependence on the substrate, stacking sequence, rate of deposition, etc.

1.4 Thin-film deposition techniques

Thin-film deposition techniques are of utmost importance which dictates the quality of thin films and their desirable properties. Before we venture into the application of reflectometry techniques for characterizing thin films, a brief discussion of the thin-film deposition techniques that are used by researchers and for commercial use is relevant and necessary. In general, thin films refer to coatings on a surface that is much smaller in thickness than the bulk material. On the lower side, graphene (a carbon monolayer) has a thickness of 0.33 nm or 3 Å [89]. On the other hand, electrodeposited metal coatings typically have a thickness in the range of microns. A material of micron thickness can be classified as a bulk material. Our target in this book is to understand emerging new phenomena at thin-film interfaces. The thin films of interest are (a) deposited on flat substrates and (b) have a thickness in the range of tens of Å to a few thousand Å.

Thin-film deposition techniques can be categorized broadly into two processes: (a) physical vapour deposition (PVD) and chemical vapour deposition (CVD). As the names suggest, CVD involves a chemical reaction, either on the substrate (probably heated) itself that releases the target atom or molecule. If the reaction takes place in a solution, the atom may precipitate and be deposited on the substrate. Electrodeposition is possibly the most widely exploited technique under CVD, where the substrate comes in the form of an electrode, on which the target atoms are deposited through an electrolysis process. For protective metallic coatings, electroplating is often used commercially. Spray pyrolysis is a CVD process where a solution is sprayed on a heated target to cause a chemical reaction among the solute components and the product is deposited in the form of a film. The spin-coating technique is another often-used process for organic films such as photoresists. In this case, a solution of the required molecules is spread near the centre of a rotating disc. The solution spreads uniformly over the disc surface due to centrifugal force. Once the solvent vaporizes, the disc is left with a thin film of the solute molecule. This technique is popular for the deposition of photoresist films. The thickness of the film depends on the speed of the rotating disc and the viscosity of the solution. Usually, for CVD films one obtains a typical thickness varying between 1 μm and 10 μm. The thickness of the films and their

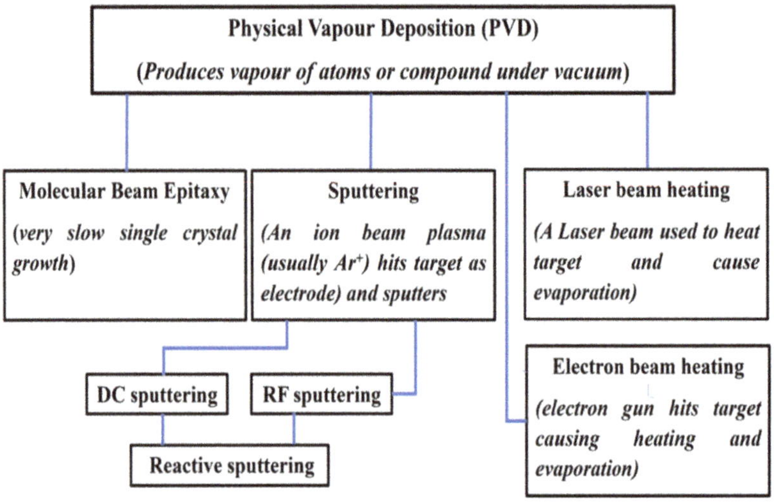

Figure 1.6. PVD techniques for thin-film deposition.

interface roughnesses are far too large in the case of CVD techniques, and they cannot be used to deposit ultra-thin films of thicknesses ranging from 10 nm to a few hundred nanometres with smooth interfaces. Of course, it needs to be mentioned that the metallo-organic CVD technique (MO-CVD) is used to grow epitaxial metal films on semiconductor surfaces in the semiconductor industry.

PVD techniques always have a target or targets of material to be deposited under a vacuum or in plasma to produce the atomic or molecular vapour to be deposited. There will be multiple targets when one plans to have layers with different components or for alloys and compounds. For the deposition of films with desirable thickness and interface properties that may be tailored, PVD techniques are preferred. Also, since PVD is usually done under a vacuum, one obtains much cleaner films in this process. There are primarily three processes associated with PVD: (i) Generation of atomic or molecular vapour at the target, (ii) transportation of the vapour to the substrate, and (iii) deposition of the atoms/molecules on the target and possible redistribution. Depending on these processes, PVD can be sub-categorized into several major techniques. Figure 1.6 lists the most commonly used PVD techniques. Depending on the way one produces the atomic/molecular vapour, the techniques can be categorized differently and are described below.

1.4.1 Resistive heating

In this case, the target and the substrate are placed inside a vacuum chamber with usually a vacuum better than 10^{-5} Torr. The target is heated in the vacuum chamber close to or above the melting point using heater coils. Since material might flow, the target is usually at the bottom of the vacuum chamber and the substrate is above it at a certain distance. The vapour plume of the target material rises in the vacuum and is deposited on the target. Since the vapour rises in form of a plume, the distance between the target and the substrate dictates the uniformity of the film thickness on the substrate: the farther the substrate, the better the uniformity. Also, the substrate should face directly at the target for better deposition. The distance between the target and substrate is usually more than a metre. On the other hand, increasing this distance makes it necessary to increase the size of the vacuum chamber and the corresponding load on the vacuum system increases. Moreover, the time required to deposit a film of a certain thickness goes up. One does not attempt to deposit films with an area larger than 100 cm^2 using this technique.

1.4.2 Electron beam heating

This process involves heating the target with an electron beam. An electron beam is produced by the thermal heating of a filament and then it is accelerated by using an approximately kilovolt field. The target is held in a crucible and the electron beam is directed to the target by the application of a magnetic field. This deposition technique also needs a high vacuum (better than 10^{-5} Torr). One may also use radio frequency (RF) or eddy current heating in the case of a conducting target. In the case of high energy, e-beam x-rays may also be emitted by the target.

Deposition systems with a heating process using heaters or e-beam are similar and have similar advantages and disadvantages. For both systems, one may use more than one target with similar heating processes. One may co-evaporate for compound films or systematically deposit periodic multiple layers or multilayers by using one target after another periodically. In the case of e-beam heating, the target is heated locally by the electrons and behaves like a point source generating a vapour plume of the target. Since the processes creates a plume of vapour, the target-to-substrate distance needs to be large and the position of the target relative to the substrate should be correct. In these heating-based deposition systems, there is usually also a substrate heating system available to heat the substrate during deposition or for *in situ* annealing post-deposition. Most importantly, there is always a quartz crystal monitor inside the vacuum chamber in the path of the plume to monitor the deposition rate. The frequency of the quartz crystal depends on its mass, which changes with the deposition of atomic vapour at a known rate. One cannot use a reactive gas in the chamber, since that will impede the vapour movement to the substrate.

1.4.3 Molecular beam epitaxy (MBE)

In principle, this technique is a part of thermal beam deposition as it follows a similar process for heating the target. However, it is worthwhile to discuss it separately, since a large number of epitaxial semiconductor layers are preferably grown by this technique. The technique is used for the growth of epitaxial films on a heated single-crystal substrate under ultra high vacuum (UHV) conditions (10^{-8} Torr to 10^{-12} Torr). The film has similar in-plane lattice parameters (within ~5%–10%) as that of the single-crystal substrate and grows in a crystalline registration with the substrate. The molecular or atomic beam has thermal energy and the growth process is much slower compared to other techniques (~1000 nm h^{-1} compared to 500 nm min^{-1} to 1000 nm min^{-1} for e-beam or sputtering). This allows the growth of crystalline layers on the substrate. The target materials are kept in effusion cells of high purity. Often MBE systems are also loaded with the RHEED technique for monitoring the layer-by-layer growth of the film. The semiconductor industry uses the technique for the growth of II–V semiconductor layers epitaxially. Prominent examples are AlGaAs on GaAs substrate and InGaAsN on GaAs. This is a special technique, used for samples of very high purity and a deposition chamber is usually dedicated to specific film deposition to avoid any contamination.

1.4.4 Magnetron sputtering

In the case of sputtering an ionized plasma is created in the vacuum chamber (usually Ar^{+}) and the target is the cathode in the system. The positive gas ion is accelerated towards the target in an electric field. The gas ion, hitting the target with high kinetic energy, ejects or sputters the atoms out of the target. The ejected atoms are transported to the substrate which is the anode in the system and are deposited. Since plasma can be large in volume, one can deposit larger-size films using the technique. Also, the target substrate distance can be much lower than that in the thermal deposition process because the sputtering takes place from the entire surface of the target, unlike e-beam heating where the melting and evaporation occur locally. Thin-film devices requiring large-area films are deposited using magnetron sputtering. For example, neutron supermirrors of the typical length of a metre and tens of centimetres in width are usually deposited by the sputtering technique. During sputtering the atoms from the target are literally *knocked out* with large kinetic energy due to momentum transfer by the impinging ion. The atoms, once deposited on the substrate, may cause redistributions of the atoms already deposited because of their higher kinetic energy. The term magnetron is used because a magnetic field is used to confine electrons near the target surface in the plasma by placing magnets in strategic locations below the target/targets. The electron path is made longer near the target surface by its helical motion in the magnetic field, thus causing more efficient ionization of the sputtering gas through collision at a lower pressure. This also confines the plasma close to the target surface.

The schematic with different components of a typical magnetron sputtering unit is shown in figure 1.7. In a typical sputtering system, the deposition chamber is first evacuated to a low base pressure ($<10^{-8}$ Torr) to reduce contamination due to background gas molecules. After achieving the base pressure, the sputtering gas (a high Z inert gas, usually Ar) is introduced into the chamber at a pressure in the range of millitorrs followed by the application of the field to create the plasma. Sputtering does not require the melting of the target for deposition, unlike the e-beam technique. This allows using a much wider range of targets for deposition. The deposition rate is associated with the sputtering yield of the target material. Depending on the energy of the impinging ion, the sputtering yield can be in the range of a fraction of unity to more than one. One can use multiple targets for sputtering several elements together to form alloys on the substrate. Since the sputtering rate is different for different atoms, depending on their mass, sputtering is not a preferred technique for use with compound targets. One can either supply power to all the targets together to allow

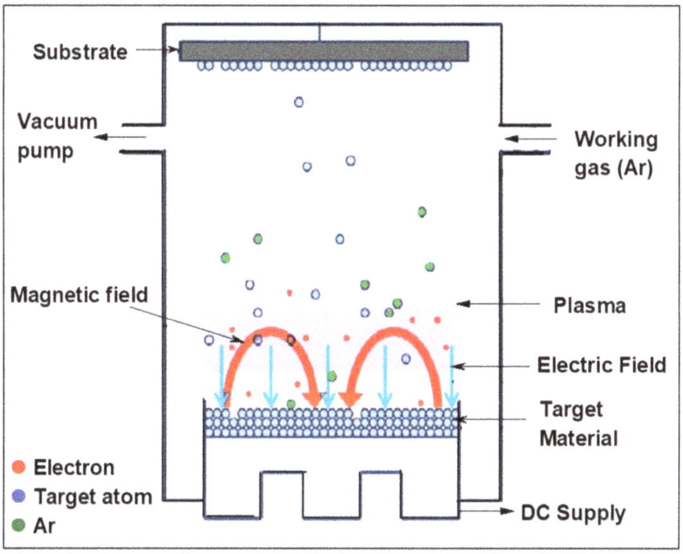

Figure 1.7. Representation of a typical magnetron sputtering system. Different components are also indicated.

mixing on the substrate or systematically switch power from one target to another for the deposition of multilayers of different elements. Using a DC sputtering system, it is not possible to deposit any insulating targets (e.g. quartz) because of the need for a very large voltage to sustain the plasma. Also due to the largely different mobility of the heavier ions compared to the electrons, there is a charge built up on the target that stops the sputtering process. RF sputtering remedies the difficulty of charge built up and helps in the sputtering of insulating materials. Brief descriptions of DC sputtering, RF sputtering, and reactive sputtering systems are described below.

A. DC sputtering

In this technique DC power is used for depositing conductive materials such as metals or transparent conductive oxides (TCOs) and it is an effective and economical choice for growing thin films of these materials. For magnetic materials (metals) such as Fe, Ni, or Co, DC magnetrons can be considered using magnets specifically selected for generating plasma within a strong magnetic field.

B. RF sputtering

In RF sputtering one applies a radio frequency voltage (typically 13.5 MHz) to the system. The oscillating voltage does not allow the charge to build on an insulating target. In the positive cycle electrons move to the cathode and in the negative cycle the sputtering is performed. RF sputtering can be performed at a lower pressure of the plasma compared to DC sputtering.

C. Reactive sputtering

This technique is used for the deposition of compounds such as oxides and nitrides. In this case, reactive gas such as oxygen and nitrogen are introduced into the sputtering chamber, together with the gas used for plasma generation. The process involves the sputtering of an element such as Fe, Ni, etc. The reactive gas forms a compound inside the chamber after a chemical reaction with the target element and the compound is deposited on the substrate. Metal nitrides such as Fe_4N and Ni_4N are examples of such nitride thin films, deposited by the technique. In this case the flow rate of the reactive gas needs careful adjustment to obtain a proper stoichiometry of the compound on the substrate.

1.4.5 Pulsed laser beam deposition (PLD)

In this case a high-power pulsed laser beam hits a target inside a vacuum chamber. This ablates the target and a plume of the target material is generated that is deposited on a substrate. This is a preferred technique for the deposition of compound films since the vapour generated in the process maintains the stoichiometry of the target compound. A large number of functional materials such as oxide ferroelectrics and high-temperature superconductors are routinely deposited using this technique. Particularly in compounds where oxygen stoichiometry is an important issue, such as oxide superconductors, this is a preferred technique. Figure 1.8 shows a schematic and photograph of a PLD unit used for growing complex oxide thin films. Usually, a KrF excimer laser is used to generate the high-power laser pulse.

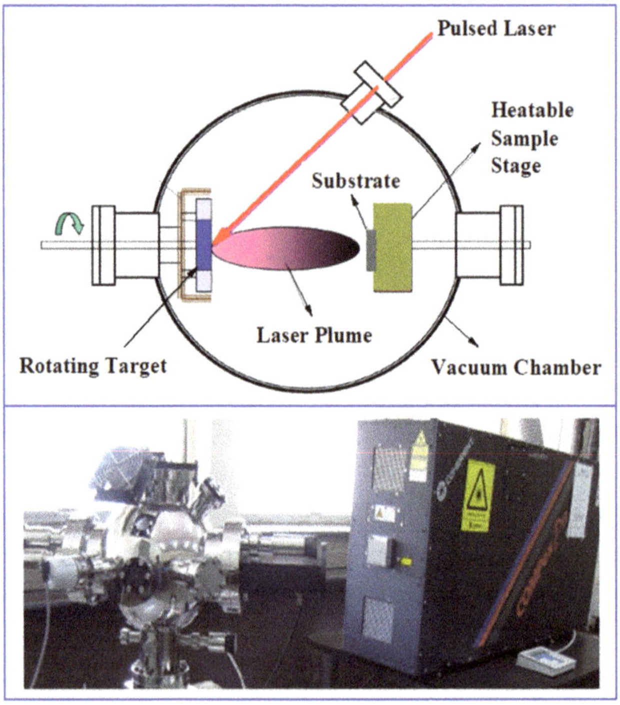

Figure 1.8. Top: Schematic of the pulsed laser deposition (PLD) system. Bottom: A photograph of the PLD chamber with a laser unit used to grow complex oxide heterostructures.

Typically the laser pulse width is in the range of tens of nanoseconds. For oxide films, in particular high T_C superconductor films, where oxygen deficiency is detrimental to the film quality, oxygen is introduced in the chamber during deposition. Also, post-deposition annealing of the films in an oxygen environment is carried out.

1.4.6 High power pulsed magnetron sputtering (HiPPMS)

This is a relatively new technique that combines the advantage of magnetron sputtering with the application of narrow pulses of high power to the system. In this process, a stream of narrow rectangular pulses of short duration is applied to the system. Since the voltage pulses are narrow (~100 μs), although the average power is comparable to DC magnetron sputtering, the peak power can be in the range of kW or MW. The pulsed power generates a very high-density plasma (~10^{19} per cubic metre) almost four orders of magnitude higher than a DC plasma. The high density and pulsed power also assist in enhancing self-sputtering by the target ions from the plasma. This process allows self-sustenance of the plasma and the requirement for Ar gas pressure goes down. The technique allows the deposition of cleaner films.

The aim of this brief description of the various physical mechanisms of thin-film deposition with a large number of deposition techniques has been to familiarize the reader with the large range of possibilities for tailoring the interface properties. One needs to make a careful choice of the materials for deposition, the technique, as well as the characterization tools for the target material, based on the above. We will demonstrate through examples the need for understanding these processes for the deposition of thin-film heterostructures with desirable properties.

1.5 Other complementary surface characterisation techniques

In addition to the high-resolution depth-dependent and non-destructive characterization tools (XRR and PNR), there are several other techniques to study the surfaces of thin films and multilayers, which use photons, ions, and electrons as surface probes. A comparison of the most common techniques is provided in table 1.2.

At present the literature on experimental and theoretical studies on emerging phenomena at the interfaces is huge. Specifically, the oxide interfaces have become the foremost in such studies and are of utmost interest to the thin-film community at the moment. The authors will focus on the experimental studies of such films using neutron (and x-ray) reflectometry in this book. It is relevant that we mention briefly the topics that will be taken up in the later chapters. This book has been organized into five chapters. The theory of neutron and x-ray reflectometry will be presented in chapter 2 together with numerical techniques that are used for the analysis of reflectometry data. The interfaces in a stack of layers of a thin film give rise to a large number of phenomena because they are physically and chemically distinct from the individual components. The translational symmetry is broken at an interface, and the composition of an interface region is different from both components on either side of the interface with different physical properties. In addition, there will be a change in

Table 1.2. Overview of techniques other than neutron and x-ray reflectometry available for the characterization of thin films.

Technique	Basic principles	Advantages and limitations
Rutherford back scattering (RBS)	Energetic ions scatter from the structure and depth-dependent structure can be determined from energy loss measurements.	Element-specific and non-destructive, but provides depth-dependent information with poor depth resolution (a few tens of nm) as compared to reflectometry techniques (sub-nm).
X-ray photoelectron spectroscopy (XPS)	Photons excite atoms, which emit an inner shell electron and provide an element-specific signal.	It is a non-destructive and element-specific technique, but no in-depth information is obtained.
X-ray florescence (XRF)	Photons excite atoms, which emit a characteristic x-ray photon.	It is non-destructive and element-specific, but no in-depth information is obtained.
Secondary ion mass spectrometry (SIMS)	Sputtered ions from the surface are analyzed by mass spectrometry.	It is an element sensitive technique, but has poor depth resolution compared to reflectometry and it is destructive.
Transmission electron microscopy (TEM)	Transmitted energetic electrons are imaged.	It is sensitive to the microscopic structure and only localized structure information is obtained. This technique is destructive and sample preparation is also laborious. It allows a visual impression of the thin films with near-atomic resolution.
Electron energy loss spectroscopy (EELS)	Measurement of loss of energy of transmitted electrons in a TEM set-up.	It is destructive and element-specific with limited depth sensitivity
Auger electron spectroscopy (AES)	Impinging electrons excite atoms from a sample, which emit an electron.	It is element-specific, but limited to only surface layers.
Scanning electron microscopy (SEM)	In this process, electron probes excite atoms of the sample and the emitted electrons are imaged.	It gives microscopic structure, symmetry, and resolution that can be varied, but with no depth resolution. Provides a visual image of the surface.
Atomic force microscopy	It works on van der Waal's force between the probe and surface.	It provides a direct image of the surface (film–air interface) with good lateral resolution. It can only probe surfaces.
Scanning tunnelling microscopy (STM)	It measures the tunnelling current between the probe and the surface of the film.	It gives a direct image of the surface with a good lateral resolution. It can only probe the surface of the sample and that of conductors only.
Ellipsometry	It is a specular reflection of light and measures thickness through a dielectric constant.	It provides the thickness and roughness of the surface layer. It has limited depth sensitivity and works well for transparent surfaces.

(*Continued*)

Table 1.2. (*Continued*)

Technique	Basic principles	Advantages and limitations
Grazing incidence x-ray diffraction (GIXRD)	X-ray diffraction from the thin surface layer.	It determines the crystal structure of the surface layer with poor depth resolution compared to reflectometry.
Magneto-optical Kerr effect (MOKE)	Rotation of the polarization of reflected light in the presence of a surface magnetic field is measured.	It is sensitive to the surface magnetic structure but has a limited depth sensitivity.
X-ray magnetic circular dichroism (XMCD)	The difference between two x-ray absorption spectra (XAS) with different polarizations (left and right circularly polarized) is taken in a magnetic field.	It is sensitive to the surface magnetic structure but has a limited depth sensitivity. It can provide information about the spin and orbital part of the magnetic moment.
X-ray resonant magnetic reflectivity (XRMR)	Soft x-ray resonant magnetic scattering offers a unique element-, site-, and valence-specific probe to study magnetic structures on the nanoscopic length scale.	Depth sensitivity is comparable to the neutron probe.

magnetism (magnetic to nonmagnetic, FM to AF phase, etc) and electrical transport properties (conductor to insulator, conductor to semiconductor, etc) at an interface.

The interface magnetism and electronic properties will add to the possibilities with respect to such physical behaviour. The interface gives rise to many interesting and unique phenomena different from the bulk of the components. We will present specific examples of emerging phenomena at the interfaces in chapter 3, in particular those studied using depth-dependent characterization, namely XRR, NR, and PNR. The examples of such phenomena have given exciting new results and have drawn the interest of researchers. In chapter 4 we will demonstrate another important aspect of reflectivity techniques called off-specular reflectivity by giving a few important results for studying the magnetic roughness and its correlation with average magnetic properties of technologically important heterostructures, e.g. thin films of 3*d* elements, RE–TM heterostructures, etc. Finally, in chapter 5 we will conclude and provide a summary of the systems discussed in this book.

References

[1] Chopra K L 1983 *Thin Film Device Applications* (New York: Springer)

[2] The Nobel Prize in Physics 2007 *The Nobel Prize* https://nobelprize.org/prizes/physics/2007/press-release/

[3] Baibich M N, Broto J M, Fert A, Van Dau F N, Petroff F, Etienne P, Creuzet G, Friederich A and Chazelas J 1988 Giant magnetoresistance of (001)Fe/(001)Cr magnetic superlattices *Phys. Rev. Lett.* **61** 2472–5

[4] Žutić I, Fabian J and Das S 2004 Sarma, spintronics: fundamentals and applications *Rev. Mod. Phys.* **76** 323–410

[5] Nakagawa N, Hwang H Y and Muller D A 2006 Why some interfaces cannot be sharp *Nat. Mater.* **5** 204–9

[6] Ohtomo A and Hwang H Y 2004 A high-mobility electron gas at the $LaAlO_3/SrTiO_3$ heterointerface *Nature* **427** 423–6

[7] Brinkman A, Huijben M, van Zalk M, Huijben J, Zeitler U, Maan J C, van der Wiel W G, Rijnders G, Blank D H A and Hilgenkamp H 2007 Magnetic effects at the interface between non-magnetic oxides *Nat. Mater.* **6** 493–6

[8] Fitzsimmons M R *et al* 2011 Upper limit to magnetism in $LaAlO_3/SrTiO_3$ heterostructures *Phys. Rev. Lett.* **107** 217201

[9] Dikin D A, Mehta M, Bark C W, Folkman C M, Eom C B and Chandrasekhar V 2011 Coexistence of superconductivity and ferromagnetism in two dimensions *Phys. Rev. Lett.* **107** 056802

[10] Hoppler J *et al* 2009 Giant superconductivity-induced modulation of the ferromagnetic magnetization in a cuprate–manganite superlattice *Nat. Mater.* **8** 315–9

[11] Satapathy D K *et al* 2012 Magnetic proximity effect in $YBa_2Cu_3O_7/La_{2/3}Ca_{1/3}MnO_3$ and $YBa_2Cu_3O_7/LaMnO_{3+\delta}$ superlattices *Phys. Rev. Lett.* **108** 197201

[12] Prajapat C L, Singh S, Paul A, Bhattacharya D, Singh M R, Mattauch S, Ravikumar G and Basu S 2016 Superconductivity-induced magnetization depletion in a ferromagnet through an insulator in a ferromagnet–insulator–superconductor hybrid oxide heterostructure *Nanoscale* **8** 10188–97

[13] Singh S, Bhatt H, Kumar Y, Prajapat C L, Satpati B, Kinane C J, Langridge S, Ravikumar G and Basu S 2020 Superconductivity-driven negative interfacial magnetization in $YBa_2Cu_3O_{7-\delta}/SrTiO_3/La_{0.67}Sr_{0.33}MnO_3$ heterostructures *Appl. Phys. Lett.* **116** 022406

[14] Nogués J and Schuller I K 1999 Exchange bias *J. Magn. Magn. Mater.* **192** 203–32

[15] Lawal A T 2020 Recent progress in graphene based polymer nanocomposites *Cogent Chem.* **6** 1833476

[16] Chadwick J 1932 Possible existence of a neutron *Nature* **129** 312–2

[17] Fermi E and Zinn W 1946 Reflection of neutrons on mirrors *Phys. Rev.* **70** 103A

[18] Felcher G P 1999 Polarized neutron reflectometry - a historical perspective *Physica* B **267–268** 154–61

[19] Parratt L G 1954 Surface studies of solids by total reflection of c-rays *Phys. Rev.* **95** 359–69

[20] Singh S, Chien T-Y, Guest J R and Fitzsimmons M R 2012 Correlation between surface rumpling and structural phase transformation in $SrTiO_3$ *Phys. Rev.* B **85** 115450

[21] Datta S and Das B 1990 Electronic analog of the electro-optic modulator *Appl. Phys. Lett.* **56** 665–7

[22] Flatté M E and Vignale G 2001 Unipolar spin diodes and transistors *Appl. Phys. Lett.* **78** 1273–5

[23] Žutić I, Fabian J and Das Sarma S 2002 Spin-polarized transport in inhomogeneous magnetic semiconductors: theory of magnetic/nonmagnetic p–n junctions *Phys. Rev. Lett.* **88** 066603

[24] van 't Erve O M J, Hanbicki A T, Holub M, Li C H, Awo-Affouda C, Thompson P E and Jonker B T 2007 Electrical injection and detection of spin-polarized carriers in silicon in a lateral transport geometry *Appl. Phys. Lett.* **91** 212109

[25] Suzuki T, Sasaki T, Oikawa T, Shiraishi M, Suzuki Y and Noguchi K 2011 Room-temperature electron spin transport in a highly doped Si channel *Appl. Phys. Express* **4** 023003

[26] Jedema F J, Filip A T and van Wees B J 2001 Electrical spin injection and accumulation at room temperature in an all-metal mesoscopic spin valve *Nature* **410** 345–8

[27] Jedema F J, Heersche H B, Filip A T, Baselmans J J A and van Wees B J 2002 Electrical detection of spin precession in a metallic mesoscopic spin valve *Nature* **416** 713–6

[28] Schleberger M, Walser P, Hunziker M and Landolt M 1999 Amorphous Fe–Si and Fe–Ge nanostructures quantitatively analyzed by x-ray-photoelectron spectroscopy *Phys. Rev.* B **60** 14360–5

[29] Anderson G W, Ma P and Norton P R 1996 The growth of Fe on sulphur passivated Ge (100): a technique for avoiding intermixing *J. Appl. Phys.* **79** 5641–3

[30] Ma P and Norton P R 1997 Growth of ultrathin Fe films on Ge(100): structure and magnetic properties *Phys. Rev.* B **56** 9881–6

[31] Freeland J W, Kodama R H, Vedpathak M, Erwin S C, Keavney D J, Winarski R, Ryan P and Rosenberg R A 2004 Induced Ge spin polarization at the Fe/Ge interface *Phys. Rev.* B **70** 033201

[32] Oyarzún S *et al* 2016 Evidence for spin-to-charge conversion by Rashba coupling in metallic states at the Fe/Ge(111) interface *Nat. Commun.* **7** 13857

[33] Singh S, Basu S, Vedpathak M, Kodama R H, Chitra R and Goud Y 2005 Ultrathin multilayer of Fe and Ge: structure and magnetic properties *Appl. Surf. Sci.* **240** 251–9

[34] Singh S, Basu S and Gupta M 2006 Structure and magnetic properties of Fe/Ge multilayer by neutron reflectometry *Physica* B **385–386** 653–5

[35] Singh S, Basu S, Gupta M, Vedpathakz M and Kodama R H 2007 Investigation of interface magnetic moment of Fe/Ge multilayer: a neutron reflectivity study *J. Appl. Phys.* **101** 033913

[36] Singh S, Basu S, Bhattacharya D and Poswal A K 2010 Physical and magnetic roughness at metal–semiconductor interface using x-ray and neutron reflectometry *J. Appl. Phys.* **107** 123903

[37] Taub A I and Fleischer R L 1989 Intermetallic compounds for high-temperature structural use *Science* **243** 616–21

[38] Singh S, Basu S, Bhatt P and Poswal A K 2009 Kinetics of alloy formation at the interfaces in a Ni–Ti multilayer: x-ray and neutron reflectometry study *Phys. Rev.* B **79** 195435

[39] Singh S, Basu S, Gupta M, Majkrzak C F and Kienzle P A 2010 Growth kinetics of intermetallic alloy phase at the interfaces of a Ni/Al multilayer using polarized neutron and x-ray reflectometry *Phys. Rev.* B **81** 235413

[40] Swain M, Singh S, Basu S and Gupta M 2013 Effect of interface morphology on intermetallics formation upon annealing of Al–Ni multilayer *J. Alloys Compd.* **576** 257–61

[41] Swain M, Singh S, Basu S, Bhattacharya D and Gupta M 2014 Identification of a kinetic length scale which dictates alloy phase composition in Ni–Al interfaces on annealing at low temperatures *J. Appl. Phys.* **116** 222208

[42] Swain M, Singh S, Basu S, Bhattacharya D, Tokas R B and Gupta M 2015 Stoichiometry dependent inter diffusion and structural evolution in Al–Ni multilayer *J. Alloys Compd.* **631** 46–51

[43] Swain M, Singh S, Bhattacharya D, Singh A, Tokas R B, Prajapat C L and Basu S 2015 Micro-structural characterization of low resistive metallic Ni germanide growth on annealing of Ni–Ge multilayer *AIP Adv.* **5** 077129

[44] Singh S, Prajapat C L, Gupta M and Basu S 2018 Room temperature superparamagnetism in ternary $(Fe_{50}Pt_{50})_{0.42}Cu_{0.58}$ phase at interfaces on annealing of $Fe_{50}Pt_{50}$/Cu multilayer *J. Magn. Magn. Mater.* **462** 58–69

[45] Basha M A, Bhatt H, Kumar Y, Prajapat C L, Gupta M, Karki V, Ghosh S K, Basu S and Singh S 2020 Evolution of structural and magnetic properties of FePtCu alloy films on annealing of FePt/Cu multilayers *Phys. Chem. Chem. Phys.* **22** 16107–16

[46] Basha M A, Bhatt H, Kumar Y, Prajapat C L, Gupta M, Karki V, Basu S and Singh S 2020 Annealing driven positive and negative exchange bias in Fe–Cu–Pt heterostructures at room temperature *J. Alloys Compd.* **815** 152640

[47] Yoo C S, Cynn H, Söderlind P and Iota V 2000 New β(fcc)-cobalt to 210 GPa *Phys. Rev. Lett.* **84** 4132–5

[48] Matar S F, Houari A and Belkhir M A 2007 *Ab initio* studies of magnetic properties of cobalt and tetracobalt nitride Co_4N *Phys. Rev.* B **75** 245109

[49] Chason E, Sheldon B W, Freund L B, Floro J A and Hearne S J 2002 Origin of compressive residual stress in polycrystalline thin films *Phys. Rev. Lett.* **88** 156103

[50] Banu N, Singh S, Satpati B, Roy A, Basu S, Chakraborty P, Movva H C P, Lauter V and Dev B N 2017 Evidence of formation of superdense nonmagnetic cobalt *Sci. Rep.* **7** 41856

[51] Kim K-J *et al* 2017 Fast domain wall motion in the vicinity of the angular momentum compensation temperature of ferrimagnets *Nat. Mater.* **16** 1187–92

[52] Finley J and Liu L 2016 Spin-orbit-torque efficiency in compensated ferrimagnetic cobalt-terbium alloys *Phys. Rev. Appl.* **6** 054001

[53] Vedmedenko E Y and Altwein D 2014 Topologically protected magnetic helix for all-spin-based applications *Phys. Rev. Lett.* **112** 017206

[54] Dzemiantsova L V, Meier G and Röhlsberger R 2015 Stabilization of magnetic helix in exchange-coupled thin films *Sci. Rep.* **5** 16153

[55] Mangin S *et al* 2014 Engineered materials for all-optical helicity-dependent magnetic switching *Nat. Mater.* **13** 286–92

[56] Fust S, Mukherjee S, Paul N, Stahn J, Kreuzpaintner W, Böni P and Paul A 2016 Realizing topological stability of magnetic helices in exchange-coupled multilayers for all-spin-based system *Sci. Rep.* **6** 33986

[57] Singh S *et al* 2019 Antisymmetric magnetoresistance and helical magnetic structure in a compensated Gd/Co multilayer *Phys. Rev.* B **100** 140405

[58] Basha M A *et al* 2020 Helical magnetic structure and exchange bias across the compensation temperature of Gd/Co multilayers *J. Appl. Phys.* **128** 103901

[59] Basha M A, Prajapat C L, Bhatt H, Kumar Y, Gupta M, Kinane C J, Cooper J F K, Langridge S, Basu S and Singh S 2020 Field dependent helical magnetic structure in a compensated Gd/Co multilayer *J. Magn. Magn. Mater.* **516** 167331

[60] Ueda K, Tabata H and Kawai T 1998 Ferromagnetism in $LaFeO_3$–$LaCrO_3$ superlattices *Science* **280** 1064–6

[61] Bhattacharya A, May S J, te Velthuis S G E, Warusawithana M, Zhai X, Jiang B, Zuo J M, Fitzsimmons M R, Bader S D and Eckstein J N 2008 Metal–insulator transition and its relation to magnetic structure in $(LaMnO_3)_{2n}/(SrMnO_3)_n$ superlattices *Phys. Rev. Lett.* **100** 257203

[62] He C *et al* 2012 Interfacial ferromagnetism and exchange bias in $CaRuO_3/CaMnO_3$ superlattices *Phys. Rev. Lett.* **109** 197202

[63] Grutter A J *et al* 2013 Interfacial ferromagnetism in $LaNiO_3/CaMnO_3$ superlattices *Phys. Rev. Lett.* **111** 087202
[64] Wang J *et al* 2003 Epitaxial $BiFeO_3$ multiferroic thin film heterostructures *Science* **299** 1719–22
[65] Ederer C and Spaldin N A 2005 Weak ferromagnetism and magnetoelectric coupling in bismuth ferrite *Phys. Rev.* B **71** 060401
[66] Yu P *et al* 2010 Interface ferromagnetism and orbital reconstruction in $BiFeO_3$–$La_{0.7}Sr_{0.3}MnO_3$ heterostructures *Phys. Rev. Lett.* **105** 027201
[67] Singh S *et al* 2014 Induced magnetization in $La_{0.7}Sr_{0.3}MnO_3/BiFeO_3$ superlattices *Phys. Rev. Lett.* **113** 047204
[68] Prajapat C L *et al* 2020 Interface-induced magnetization and exchange bias in LSMO/BFO multiferroic heterostructures *ACS Appl. Electron. Mater.* **2** 2629–37
[69] Singh S, Xiong J, Chen A P, Fitzsimmons M R and Jia Q X 2015 Field-dependent magnetization of $BiFeO_3$ in an ultrathin $La_{0.7}Sr_{0.3}MnO_3/BiFeO_3$ superlattice *Phys. Rev.* B **92** 224405
[70] Jain P *et al* 2015 Synthetic magnetoelectric coupling in a nanocomposite multiferroic *Sci. Rep.* **5** 9089
[71] Chakhalian J, Freeland J W, Habermeier H U, Cristiani G, Khaliullin G, van Veenendaal M and Keimer B 2007 Orbital reconstruction and covalent bonding at an oxide interface *Science* **318** 1114–7
[72] Stahn J *et al* 2005 Magnetic proximity effect in perovskite superconductor/ferromagnet multilayers *Phys. Rev.* B **71** 140509
[73] Paull O H C, Pan A V, Causer G L, Fedoseev S A, Jones A, Liu X, Rosenfeld A and Klose F 2018 Field dependence of the ferromagnetic/superconducting proximity effect in a YBCO/STO/LCMO multilayer *Nanoscale* **10** 18995–9003
[74] Kumar Y, Bhatt H, Prajapat C L, Singh A P, Singh F, Kinane C J, Langridge S, Basu S and Singh S 2021 Suppression of the superconducting proximity effect in ferromagnetic-superconducting oxide heterostructures with ion-irradiation *J. Appl. Phys.* **129** 163902
[75] Parkin S S P, Mansour A and Felcher G P 1991 Antiferromagnetic interlayer exchange coupling in sputtered Fe/Cr multilayers: dependence on number of Fe layers *Appl. Phys. Lett.* **58** 1473–5
[76] Schreyer A, Bröhl K, Ankner J F, Majkrzak C F, Zeidler T, Bödeker P, Metoki N and Zabel H 1993 Oscillatory exchange coupling in Co/Cu(111) superlattices *Phys. Rev.* B **47** 15334–7
[77] Schreyer A, Ankner J F, Zeidler T, Zabel H, Schäfer M, Wolf J A, Grünberg P and Majkrzak C F 1995 Noncollinear and collinear magnetic structures in exchange coupled Fe/Cr(001) superlattices *Phys. Rev.* B **52** 16066–85
[78] Borchers J A, Dura J A, Unguris J, Tulchinsky D, Kelley M H, Majkrzak C F, Hsu S Y, Loloee R, Pratt W P and Bass J 1999 Observation of antiparallel magnetic order in weakly coupled Co/Cu multilayers *Phys. Rev. Lett.* **82** 2796–9
[79] Langridge S, Schmalian J, Marrows C H, Dekadjevi D T and Hickey B J 2000 Quantification of magnetic domain disorder and correlations in antiferromagnetically coupled multilayers by neutron reflectometry *Phys. Rev. Lett.* **85** 4964–7
[80] Ball A R, Fredrikze H, van der Zaag P J, Jungblut R, Reinders A, van der Graaf A and Rekveldt M T 1995 Neutron reflectometry on an exchange biased $Ni_{80}Fe_{20}/Fe_{50}Mn_{50}$ bilayer *J. Magn. Magn. Mater.* **148** 46–8

[81] te Velthuis S G E, Felcher G P, Jiang J S, Inomata A, Nelson C S, Berger A and Bader S D 1999 Magnetic configurations in exchange-biased double superlattices *Appl. Phys. Lett.* **75** 4174–6

[82] Fitzsimmons M R, Yashar P, Leighton C, Schuller I K, Nogués J, Majkrzak C F and Dura J A 2000 Asymmetric magnetization reversal in exchange-biased hysteresis loops *Phys. Rev. Lett.* **84** 3986–9

[83] Paul A, Schneider C M and Stahn J 2007 Reversal mechanism and suppression of training in an exchange-coupled system *Phys. Rev.* B **76** 184424

[84] Radu F, Etzkorn M, Siebrecht R, Schmitte T, Westerholt K and Zabel H 2003 Interfacial domain formation during magnetization reversal in exchange-biased CoO/Co bilayers *Phys. Rev.* B **67** 134409

[85] Gibert M, Zubko P, Scherwitzl R, Íñiguez J and Triscone J-M 2012 Exchange bias in $LaNiO_3$–$LaMnO_3$ superlattices *Nat. Mater.* **11** 195–8

[86] Bhatt H, Kumar Y, Prajapat C L, Kinane C J, Caruana A, Langridge S, Basu S and Singh S 2020 Emergent interfacial ferromagnetism and exchange bias effect in paramagnetic/ferromagnetic oxide heterostructures *Adv. Mater. Interfaces* **7** 2001172

[87] Singh S, Prajapat C L, Bhattacharya D, Ghosh S K, Gonal M R and Basu S 2016 Antiferromagnetic coupling between surface and bulk magnetization and anomalous magnetic transport in electro-deposited cobalt film *RSC Adv.* **6** 34641–9

[88] Katmis F *et al* 2016 A high-temperature ferromagnetic topological insulating phase by proximity coupling *Nature* **533** 513–6

[89] Novoselov K S, Geim A K, Morozov S V, Jiang D, Zhang Y, Dubonos S V, Grigorieva I V and Firsov A A 2004 Electric field effect in atomically thin carbon films *Science* **306** 666–9

IOP Publishing

Neutron and X-ray Reflectometry
Emerging phenomena at heterostructure interfaces
Saibal Basu and Surendra Singh

Chapter 2

Theory of neutron reflectometry

2.1 Introduction

Neutron reflectometry (NR) as a technique comprises the measurement of a reflected beam of neutrons from the surface of a thin film that can be a single layer or multiple layers [1–8]. Since the refractive index of neutrons for most materials is marginally less than one, the reflection takes place at 'near-grazing' incidence and needs a finely collimated neutron beam of a few arc minutes. The theory of neutron reflectometry percolates from the theory of x-ray reflectometry (XRR) [4, 9] or more generally from the theory of reflection of electromagnetic waves [10]. Both NR and XRR are carried out either in a specular mode where the angle of reflection is equal to the angle of incidence and Snell's law is valid, or in an off-specular mode when the equality of the angle of incidence to the angle of reflection is violated [4]. Off-specular scattering follows from diffuse surface reflection [4, 9]. We will introduce both these techniques in this chapter. Figure 2.1 shows the geometry for specular reflection from a film plane.

For specular reflectivity, the angle of incidence $\theta_{\rm i} = \theta_{\rm f} = \theta$ and the momentum transfer vector is along the normal to the surface of the film, whose magnitude is given by $|\ \overline{Q}\ | = \overline{K}_{\rm i} - \overline{K}_{\rm f} = Q_z = \frac{4\pi}{\lambda}\sin\theta$, where λ is the wavelength of the incident radiation. As the angle of incidence increases, the momentum transfer also increases along the normal, in the case of specular reflectometry. Usually, the specular reflectivity data are collected as a function of the momentum transfer, Q_Z. One can extract information along the depth of the film with a resolution in sub-nanometre scale with neutron (and x-ray) reflectometry, which makes the technique a useful tool for characterizing thin films. In the case of off-specular scattering, the incident angle and the angle of reflection are not equal and the reflection geometry in reciprocal space is shown in figure 2.2. In this case the momentum transfer Q is not along the normal to the plane and has a component Q_x along the plane. Measurement of off-specular neutron (or x-ray) scattering allows one to obtain an in-plane height–height correlation function (interface morphology).

doi:10.1088/978-0-7503-4695-5ch2

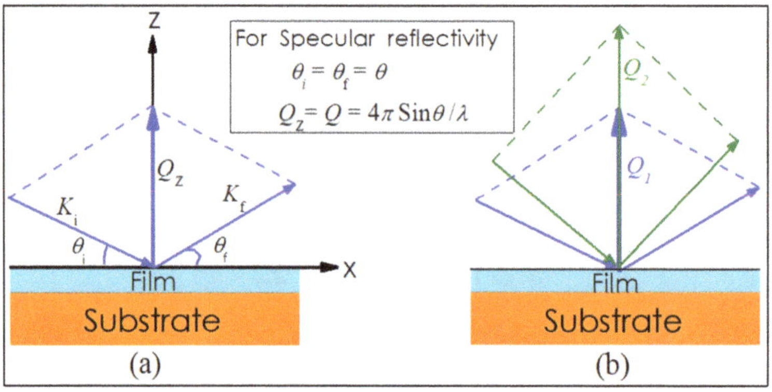

Figure 2.1. (a) Geometry of specular reflection where $\theta_i = \theta_f = \theta$ and the momentum transfer vector $\overline{Q} = \overline{K_i} - \overline{K_f} = \frac{4\pi}{\lambda}\sin\theta$. (b) As the angle of incidence increases, the momentum transfer increases ($Q_2 > Q_1$).

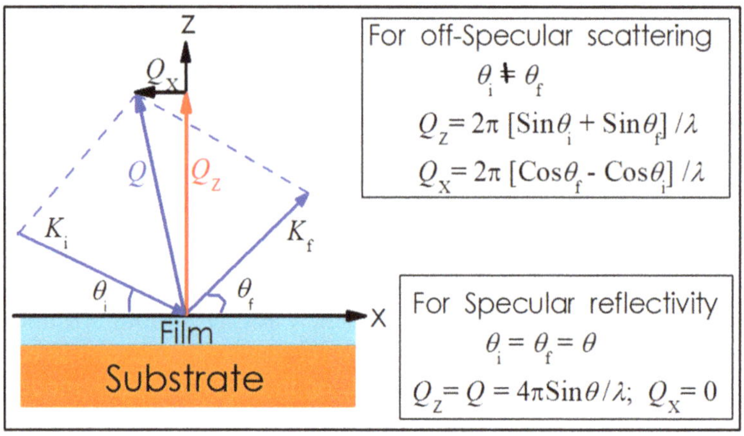

Figure 2.2. Geometry for off-specular scattering where the angle of incidence is not equal to the angle of reflection. There is an in-plane component of wavevector transfer (Q_x) in addition to the normal component (Q_z). The values of these components are also shown in the figure.

Here we will attempt to discuss the theoretical treatment of specular NR and its important extension, polarized neutron reflectometry (PNR) together with XRR. There are several books and review articles available on these techniques [1–5]. Treatment of off-specular scattering will also be presented for quantifying *true* roughness through the height–height correlation function and detailed roughness morphology of the interfaces [9]. Both the techniques, PNR and XRR have a very similar theoretical foundation. Usually, one carries out XRR to obtain the details of a layer structure (density, thickness, and interface roughness) of a heterostructure immediately after deposition. XRR as a technique is normally available at the laboratories and is used to characterize the thin film samples, before one may take them to a neutron source for further studies. The theory of off-specular neutron and x-ray scattering is also included here for studies regarding the microscopic characterization of interfaces. NR and XRR primarily differ in their interaction potential with the matter (medium). PNR is an important extension of NR where the interaction between the neutron magnetic moment and magnetism in the thin film sample comes into play, in addition to nuclear potential, for neutrons. Most of the studies that are presented here have used PNR as a primary technique, often in conjunction with XRR.

2.2 Optical theory of reflection

The reflectometry technique deals with the refractive index of the medium (the thin film here) with respect to the radiation (x-ray or neutron). It depends on the potential seen by the radiation while incident on the medium. The medium presents an average potential that depends on the respective scattering length density (SLD), i.e. electron SLD (ESLD) for x-ray and nuclear SLD (NSLD) for neutron. The potential for unpolarized neutrons can be written as

$$V(r) = \frac{2\pi\hbar^2}{m} N b_{\mathrm{coh}} = \frac{2\pi\hbar^2}{m}\rho, \tag{2.1}$$

where N and b_{coh} are the number density (nuclei/cc) of the scatterers in the medium and the coherent scattering length of each scatterer, respectively. $\hbar$ is $h/2\pi$ (h is Planck's constant) and m is the mass of the neutron. The product $\rho = Nb_{\mathrm{coh}}$ is the NSLD of the medium that can be extracted from reflectometry data. If the medium comprises several isotopes of an element i and several such elements j then: $Nb_{\mathrm{coh}} \rightarrow \sum_j \sum_i N_{ij} b_{\mathrm{coh}}^{ij}$.

Unlike x-rays, isotopes of an element have different coherent scattering lengths for thermal neutrons. Figure 2.3 shows the schematic of the nuclear potential (V) seen by a neutron of energy E_{n}. The treatment assumes that the reflecting medium behaves like a homogeneous medium with an average potential. This is unlike neutron diffraction experiments for a crystallographic structure that sees the atomic structure of the

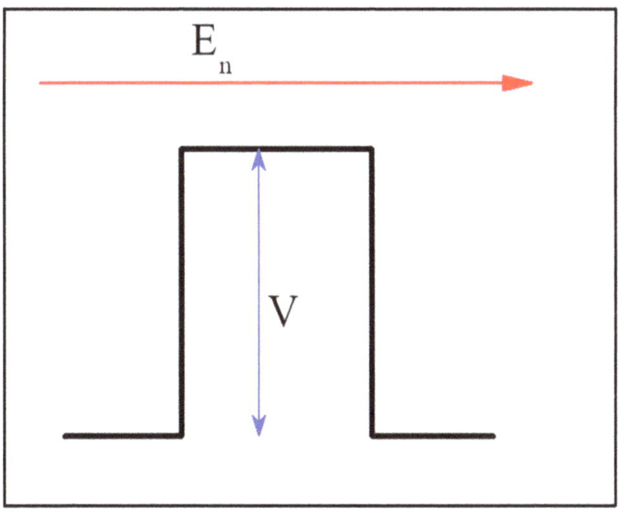

Figure 2.3. The nuclear potential V (given by equation (2.1)) experienced by a neutron in the medium. E_{n} is the energy of the neutron.

medium. The justification for treating the reflecting medium as homogeneous lies in the fact that neutron reflectometry experiments are done at a lower momentum transfer (Q) range, typically from 0.001 $Å^{-1}$ to 0.1 $Å^{-1}$. In this range of momentum transfer, the neutron wave's spatial resolution is much larger than the atomic distance in a medium and one may treat the medium as a homogeneous one. This is similar to the propagation of an *optical wave* in a medium which is described by a refractive index (n), derived from the *optical density* of the medium.

The apparent energy of the neutron in the medium is given by $E_1 = E - V$. If the wave vector of the incident neutron is K and it is K_1 in the medium. In the non-relativistic limit, the energy of the neutron in the medium is given by

$$E_1 = \frac{\hbar^2 K_1^2}{2m} = E_n - V = \frac{\hbar^2 K^2}{2m} - V. \tag{2.2}$$

The wave vector of the incident neutron K changes to $\overline{K_1}$ under influence of the nuclear potential. This allows us to correlate the refractive index of the medium, n, to the NSLD (ρ) of the medium as given in

$$n^2 = \frac{K_1^2}{K^2} = 1 - \frac{4\pi\rho}{K^2} = 1 - \frac{\lambda^2}{\pi}\rho. \tag{2.3}$$

The refractive index is given by

$$n = 1 - \frac{\lambda^2}{2\pi}\rho = 1 - \delta. \tag{2.4}$$

For most of the materials, the refractive index for neutrons is marginally less than unity with δ typically in the range of 10^{-6} and reflection of neutrons takes place at grazing incidence, as stated earlier. Since the refractive index is less than unity, there will be a total external reflection for neutrons unlike optical rays (refractive index in the medium > 1) where total internal reflection occurs. On including absorption, one can write a generalized expression for the refractive index of neutrons:

$$n = 1 - \delta + i\beta \tag{2.5a}$$

where β is related to the linear absorption coefficient for neutrons in the medium. It needs to be mentioned that neutrons usually have a very small value of absorption coefficient ($\beta = 0.0$ for most of the elements for thermal neutron) compared to x-rays and they can penetrate very deep (several centimetres!) in most materials. This is an advantage of using neutrons to study bulk materials or buried interfaces. For neutron one can write as follows:

$$\delta_{\text{neutron}} = \frac{N\lambda^2}{2\pi}\left[b_{\text{coh}} + b_{\text{disp}}\right]; \text{ and } \beta_{\text{neutron}} = \frac{N\lambda^2}{2\pi} \mid b_{\text{abs}} \mid, \tag{2.5b}$$

where b_{disp} is a dispersion term that takes care of variation of b_{coh} of the scatterer with the energy of neutrons and b_{abs} is the absorption length of a scatterer for neutrons. The corresponding expressions for x-rays are given for comparison:

$$n_{\text{x-ray}} = 1 - \frac{\lambda^2}{2\pi}N_e r_0 = 1 - \frac{\lambda^2}{2\pi}\rho_e, \tag{2.6a}$$

where N_e is the electron number density of the medium and r_0 is the classical electron radius (= 2.818 fm). λ is the wavelength of the incident x-ray beam. $\rho_e = N_e r_0$ is the electron scattering length density (ESLD):

$$\delta_{\text{x-ray}} = \frac{N_e r_0 \lambda^2}{2\pi}\left[f_0 + f_{\text{disp}}\right] \text{ and } \beta_{\text{x-ray}} = \frac{N_e \lambda^2}{2\pi}\left|f_{\text{abs}}\right|, \tag{2.6b}$$

where $f_0 \sim Z$, the atomic number for a monatomic medium. Translation to other forms of polyatomic scatterers is straightforward. f_{disp} is the dispersion term of the scatterer for x-rays and f_{abs} is the x-ray absorption length of the scatterer. The x-rays experience the 'optical potential' due to the electron density in the medium similar to what is shown in figure 2.3.

It is evident from the above expressions that apart from the difference in the potential seen by the neutrons and x-rays the theory of the techniques are similar and they are complementary to each other. Both the experimental techniques require the preparation of a highly collimated beam that impinges on the sample at grazing incidence (usually at a few arc minutes). This requires an elaborate arrangement for placing the thin film sample in the beam for measuring the reflectivity profile. XRR can be done on laboratory sources and is more easily accessible. NR is done at reactors or in spallation neutron sources and one needs to go to a large facility. Since the reflectivity is measured as a function of momentum transfer Q, we use a highly collimated monochromatic neutron beam in a reactor. In a pulsed spallation source time-of-flight (TOF) is the preferred method of data collection. Examples of data collected using both techniques will be discussed later.

One can calculate the refractive index (n) for a medium from the above potential. At any interface, wherever there is a refractive index mismatch it causes reflection and refraction of the incident beam. Neutron (and x-ray) reflectometry consists of measuring the reflected intensity at a film–air interface (surface). The film may consist of single layer (thickness ~ a few tens of nm) or layers of different materials (multilayer) with different refractive indices on a substrate. For a neutron wave incident at grazing incidence on a surface, there will be a reflected beam at the interface together with refraction and transmission of the beam in the medium. The propagation of the beam is schematically shown in figure 2.4, where r and t are the reflected and transmitted amplitudes. One measures the reflected intensity ($= r^* r$) in a reflectivity experiment. Following Snell's law, in the case of specular reflectivity, the incident angle is equal to the angle of reflection ($\theta_i = \theta_f$). Thus measuring specular reflectivity as a function of wavevector transfer (Q) perpendicular to the surface of the film provides a detailed depth-dependent NSLD (layer structure) of the film.

For off-specular reflection, this equality is not maintained and there is a component of wave vector transfer along the plane of the film (shown in figure 2.2). Off-specular reflectometry is used to find out the in-plane height–height correlation function at an interface [4, 9]. Off-specular scattering allows one to quantify roughness at the interface and separate the effect of 'inter-mixing' from true roughness. We will discuss the treatment of both specular reflectivity and off-specular scattering in this chapter. For simplicity, we will first describe the NR assuming only nuclear potential which is termed unpolarized NR.

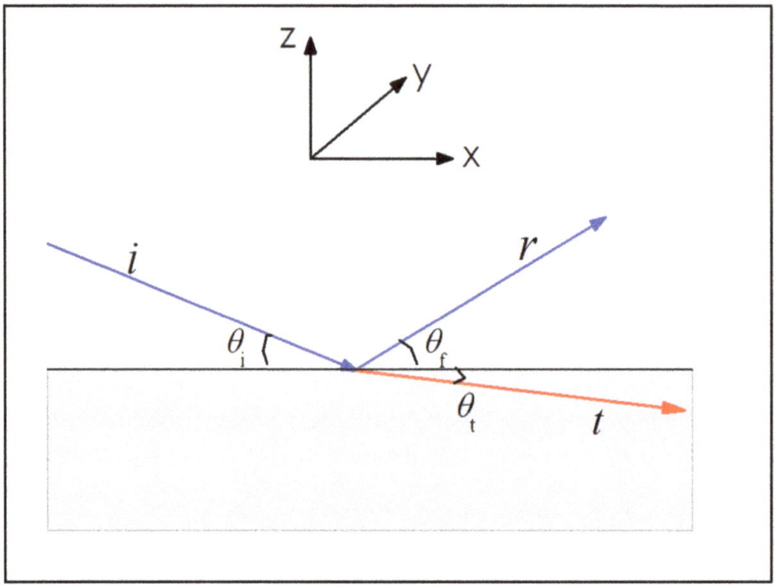

Figure 2.4. The schematic for the reflection and transmission of the neutron wave at an interface between two media. The incident beam amplitude (*i*) gets reflected (*r*) from the surface and a part of the beam gets transmitted in the medium (*t*). The plane of reflection is the (*z*–*x*) plane in the diagram. The thin film surface is the (*x*–*y*) plane and *z* is the depth in the reflecting thin film medium. For specular reflection $\theta_i = \theta_f$.

2.3 Specular NR using unpolarized neutrons

It is evident from equation (2.1) to equation (2.5b) that unpolarized NR occurs due to the interaction of neutron nuclei and it connects to the NSLD of the reflecting medium, which is quite similar to the connection of XRR to the ESLD of the medium. Both techniques relate to the physical density through NSLD (from NR) and ESLD (from XRR). It is evident from figure 2.1 that the momentum transfer Q is along the normal to the reflecting medium for specular reflectivity. One can obtain depth-resolved density information of the heterostructure from the specular reflectivity data with sub-nanometre resolution. Specular neutron reflectivity in the unpolarized mode determines the structural parameters of thin films, namely individual layer thickness, interface roughness, and the NSLD of the layers. In non-relativistic quantum mechanics, the motion of a neutron of mass, m, and energy, E_n, is given by the Schrödinger equation for probability amplitude, ψ [6]:

$$-\frac{\hbar^2}{2m}\nabla^2\psi + V\psi = E_n\psi. \tag{2.7}$$

We consider the reflection of a matter-wave represented by wave amplitude, ψ at a planar stratified boundary between two media. The potential, V, is the effective potential, representing the interaction between the wave and the medium. The boundary is characterized by *optical potentials* V_1 and V_2 for the two media. We have already stated the general forms of the refractive index for neutrons and x-rays, in equation (2.5a) and equation (2.6a), respectively. If we consider that z–x is the plane of reflection with the interface lying in the x–y plane and the stratification along the z-direction (figure 2.4), then the potential depends only on the z variable. Since the plane of reflection is the z–x plane, we may exclude the dependence of the wavefunction ψ on the y variable. If the incident wave is a plane wave, we may separate the variables z and x and the wavefunction is given by $\psi(z, x) = e^{iKx}\psi(z)$. Equation (2.7) reduces to a one-dimensional linear differential equation in variable z alone,

$$\frac{d^2\psi(z)}{dz^2} + Q_z^2\psi(z) = 0 \quad \text{with} \quad Q_z^2 = \frac{2m}{\hbar^2}[E_\text{n} - V(z)] - K^2, \tag{2.8a}$$

where Q_z is the same wave-vector as shown in figure 2.1, normal to the plane of the sample, and is along the z-direction. K is the x-component of the wave vector if z–x is the plane of reflection. We present the case for the propagation of x-rays in a medium in the box below, for the sake of comparison.

Going back to the tenets of electromagnetic waves [7, 10], let us consider the Maxwell's equation that governs the propagation of an electromagnetic wave of angular frequency ω. The wave equation follows the two curl equations relating electric field E and magnetic field B. The time variation of the electric field is given by $e^{i\omega t}$. The Maxwell's equations in SI units are

$$\nabla \times \overline{E} = i\omega\overline{B} \tag{2.8b}$$

and

$$\nabla \times \overline{B} = i\epsilon\frac{\omega}{c^2}\overline{E}, \tag{2.8c}$$

where ϵ is the dielectric constant. If we consider the electromagnetic wave propagating along the z-direction with x–z as the reflecting plane, then the electric field is $\overline{E} = \left(0, E_y, 0\right)$. For planar stratification the dielectric constant is a function of the spatial variable z only. We can write the y component of electric field

$$E_y(x, z, t) = e^{i(Kx-\omega t)}E(z). \tag{2.8d}$$

Here $E(z)$ satisfies a linear differential equation derived from equations (2.8c) and (2.8d):

$$\frac{d^2E}{dz^2} + Q^2E = 0\text{:}\quad Q^2 = \epsilon\frac{\omega^2}{c^2} - K^2 = k^2. \tag{2.8e}$$

The equations for the wave function ψ in the case of neutrons and the electric field E for electromagnetic waves follow the same form of the linear differential equation. In equations (2.8a) and (2.8e) K is the x-component of the wavevector in either medium and is a constant of motion. The laws of reflection and refraction follow from this invariance. There is no transverse component (normal to the z-direction) of *force* due to the optical potential, V, since $\frac{\partial V}{\partial x} = \frac{\partial V}{\partial y} = 0$. The angle of incidence, angle of reflection, and angle of refraction are θ_i, θ_f, and θ_t, respectively, as shown in figure 2.4. The x-components of the corresponding wavevectors are equal, given by $k_{1x}^\text{i} = k_{1x}^\text{f} = k_{2x}^\text{f}$, where k_i is the wavevectors in the corresponding media 1 and 2

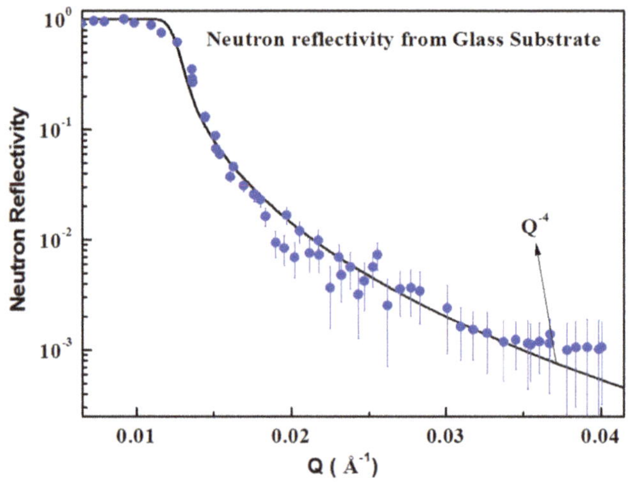

Figure 2.5. The measured neutron reflectivity from a polished glass surface along with fitted Fresnel reflectivity.

between which the reflection/transmission is taking place. In terms of the angles defined in figure 2.4:

$$k_1\cos\theta_i = k_1\cos\theta_f = k_2\cos\theta_t, \tag{2.9}$$

where

$$k_i^2 = K^2 + Q_i^2 = \frac{2m}{\hbar^2}\ [E_n - V_i], \tag{2.10}$$

where Q_i is the wavevector in the z-direction in mediums 1 and 2 and K is the in-plane component of the wavevector. A comparison of equations (2.10) and (2.8e) indicates that there is a one-to-one correspondence between the propagation of the matter-wave, obeying Schrödinger's equation, and the electromagnetic wave, following Maxwell's equations.

2.3.1 Fresnel reflectivity of an ideally flat surface

Let us consider an ideally flat surface with a sharp boundary as shown in figure 2.4. In this figure a neutron beam impinges on the surface from a vacuum at a glancing angle θ_i, and gets reflected at an angle θ_f (= θ_i for specular reflectivity) and the transmitted beam propagates at an angle θ_t in the second medium. The wave function in a vacuum (medium 1) consists of the sum of the incident wave amplitude and the reflected amplitude. We assumed the normal to the plane in the z-direction and the reflecting medium is in the x–y plane. The wavefunction across the interface is two media given by (normalized to incident amplitude as unity)

$$e^{iQ_1z} + re^{-iQ_1z}\ \text{(vacuum or air, medium 1)} \leftarrow \psi(z) \rightarrow te^{iQ_2Z}\ \text{(medium 2)}, \tag{2.11}$$

where Q_1 and Q_2 are the z-component of the wavevector transfer in mediums 1 and 2. If we consider the continuity of the wavefunction $\psi(z)$ and its derivative $d\psi/dz$ at the interface, $z = 0$, we obtain the following from the above expressions:

$$1 + r = t \text{ and } Q_1(1 - r) = Q_2 t. \tag{2.12}$$

These expressions give the expression for reflectivity and transmission amplitudes as follows:

$$r = \frac{Q_1 - Q_2}{Q_1 + Q_2} \text{ and } t = \frac{2Q_1}{Q_1 + Q_2}. \tag{2.13}$$

In this equation $Q_1 = \frac{2\pi}{\lambda}\sin\theta_i$ and $Q_2^2 = Q_1^2 - 4\pi\rho b_{\text{coh}}$. At grazing incidence $\theta \to 0$ and we can express a critical angle for the total external reflection of neutrons, related to scattering length density in the medium:

$$\theta_c = \sqrt{2(1 - n)} = \sqrt{\frac{\rho b_{\text{coh}}}{\pi}}\lambda. \tag{2.14}$$

The critical angle θ_c increases linearly with the wavelength of the neutrons and also increases with an increase in nuclear scattering length density. The Fresnel reflectivity for an ideally flat surface at a glancing angle θ_i is given by

$$R_F = r^*r = \left| \frac{\sin\theta_i - \sqrt{\sin^2\theta_i - \sin^2\theta_c}}{\sin\theta_i + \sqrt{\sin^2\theta_i - \sin^2\theta_c}} \right|^2. \tag{2.15}$$

When $\theta_i < \theta_c$ the reflection amplitude r is a complex number and the Fresnel reflectivity is unity for the angle of incidence below the critical angle. It follows from equation (2.15) that for wavevector $Q \gg Q_c$ the reflected intensity falls as Q^{-4}. The typical neutron reflectivity data (symbol) from a polished glass substrate is shown in figure 2.5, along with the fit (solid line) to Fresnel reflectivity.

The Q^{-4} drop in intensity of the reflectivity pattern makes measurement a challenging task at relatively larger angles of incidence (larger Q) beyond the critical angle. It also needs to be mentioned that none of the interfaces, in reality, are ideally flat, the fundamental assumption in the above derivation. A reflecting surface always has an associated 'surface roughness' that needs to be accounted for, in reflectivity formalism. Figure 2.6(a) shows an example of a rough surface with a height–height distribution as a Gaussian function. The FWHM of the Gaussian distribution is termed the surface roughness (σ). An atomic force microscopy image of a Ni surface, deposited by sputtering is shown in figure 2.6(b).

The role of the roughness is to make the surface/interface 'fuzzy' or smear it out by causing fluctuation around an average interface height (figure 2.6(a)). Drawing similarity from the effect of thermal vibrations on the intensity of Bragg peaks, the roughness contribution to reflectivity from a film with definite roughness was *heuristically* introduced as a 'Debye–Waller like' factor [11] that modifies the Fresnel reflectivity (R_F), due to a smooth interface, with an exponential factor that accounts for the roughness:

$$R = R_F e^{-Q_1^2\sigma^2/2}, \tag{2.16a}$$

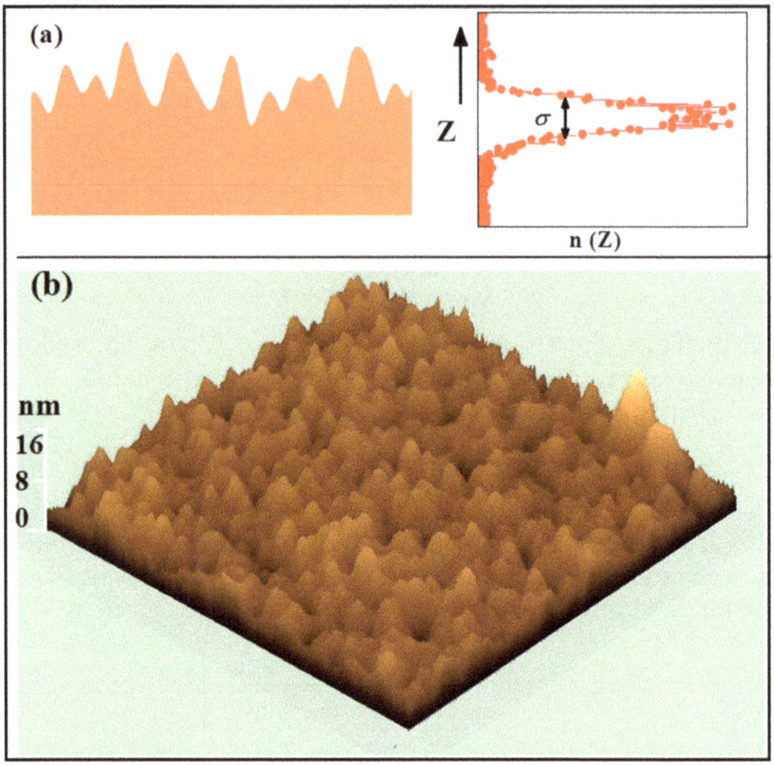

Figure 2.6. (a) Schematic of the interface roughness as fluctuation around a mean surface and the roughness parameter, σ is the FWHM of the Gaussian variation of height (z) distribution. (b) atomic force microscopy image of the surface of a Ni film deposited by sputtering.

where σ is the roughness parameter. Apart from depth-resolved structural information, we also estimate the roughness parameter from fits to neutron reflectivity data. It was Névot and Croce [12] who analytically arrived at the following expression for an interface for short-length scale roughness $Q_z\sigma \ll 1$:

$$R = R_F e^{-Q_1 Q_2 \sigma^2/2}. \tag{2.16b}$$

Since Q_1 and Q_2 are close in the kinematical limit, both equations (2.16a) and (2.16b) provide roughness parameters that are nearly the same for given reflectometry data. In the case of multiple layers or 'multilayers', the roughness parameter for an interface that one obtains from specular reflectometry is a convolution of 'interface mixing' of components as well as 'height fluctuation', since both of them tend to make the interface broader and fuzzy. It is not possible to separate these contributions from specular reflectometry data only. The height fluctuation or *true* roughness can be quantified in terms of a height–height correlation function that is obtained from off-specular reflectivity, to be discussed later in this chapter. A comparison of a simulated specular unpolarized NR pattern for a smooth (roughness ~1 Å) and a rough (roughness ~15 Å) Si substrate is depicted in figure 2.7. The inset shows the schematic of the smooth and rough Si substrate. It is evident from figure 2.7 that the reflectivity of the rough film falls rapidly at a higher Q.

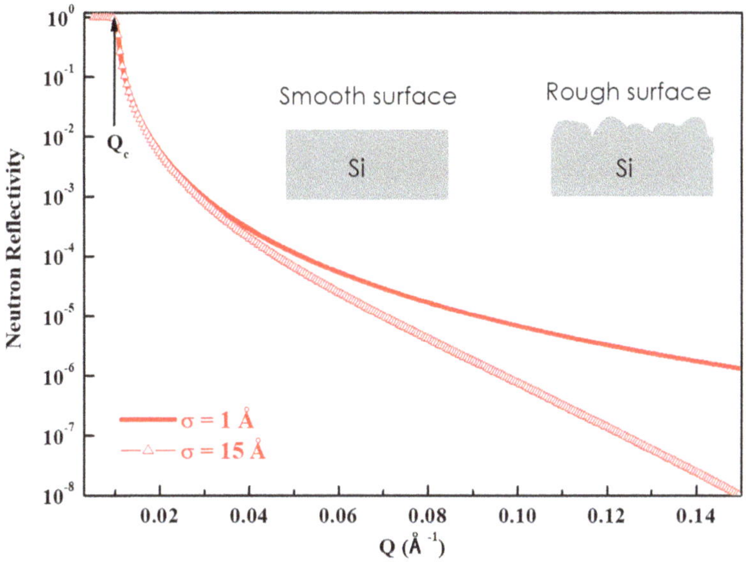

Figure 2.7. Specular NR pattern from Si substrate with negligible roughness of 1 Å (solid line) and a larger roughness of 15 Å (open triangle with line). The inset shows the schematic of the films.

2.3.2 Reflection from a film of finite thickness and a periodic multilayer

In the previous section we derived the reflectivity from an interface of a medium with infinite thickness at the film–air interface. In the case of the finite thickness of a film, the neutron beam is reflected from the surface (film–air interface), as well as from the bottom of the film (substrate–film interface) as shown in figure 2.8(a). Interference between the two reflected amplitudes at the film–air interface (A) and the transmitted amplitude (T) reflected at the film–substrate interface (B) causes interference fringes in the reflected beam known as 'Kiessig oscillations'. Such oscillations can be detected in both NR and XRR patterns. Figure 2.8(b) depicts a thin film sample with a single film of thickness L grown on the substrate, with the NSLD depth profile and derivative of NSLD profile (which is a delta function at two interfaces A and B) with respect to thickness (z-axis). Figure 2.8(c) shows the experimentally observed Kiessig oscillations in the NR pattern from an iron phthalocyanine (FePc) film along with the fit that gave a film thickness of ~1000 Å from the periodicity of the oscillations. In general, the reflected intensity can be represented as the Fourier transform of the derivative of a scattering length density (NSLD for neutron and ESLD for x-ray) profile $\rho(z)$ multiplied by the Fresnel reflectivity [7]. In the case of a layered structure along the z direction, the one-dimensional expression is given by [7]

$$R = R_{\mathrm{F}} \left| \int \frac{d\rho(z)}{dz} e^{iQ_z z} dz \right|^2. \tag{2.17}$$

For a film of thickness L with a sharp step-like density function, $\frac{d\rho(z)}{dz}$ (see figure 2.8(b)) in the above equation gives two δ-functions at $z = 0$ (film–air interface)

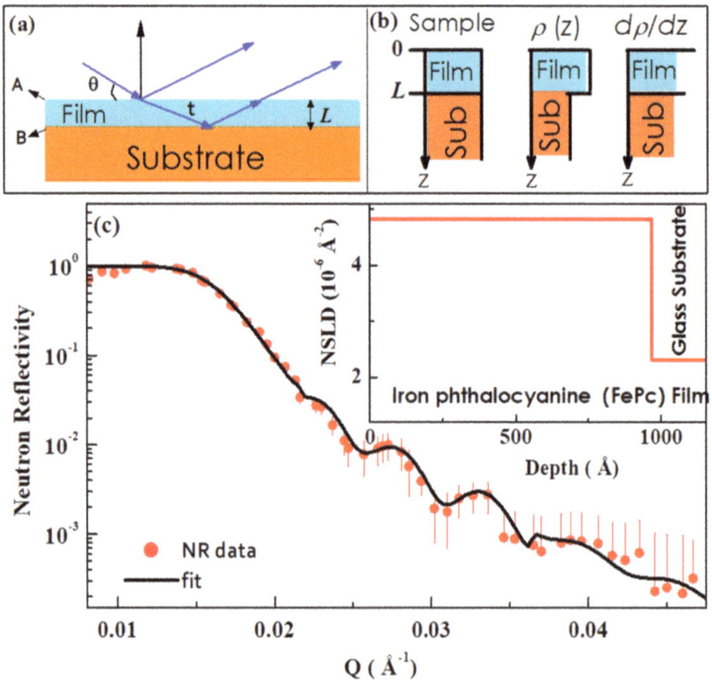

Figure 2.8. (a) Schematic of the reflected amplitudes at the top of a film at the film–air (A) interface and bottom from substrate–film (B) interface. Interference between these two waves causes oscillations in the reflected intensity, known as Kiessig oscillations. (b) Schematic of a single thin film grown on a substrate, its NSLD depth profile and derivative of NSLD profile. (c) The experimental Kiessig in NR (closed circles) pattern from an iron phthalocyanine (FePc) film of thickness ~1000 Å grown on a glass substrate and the corresponding fit (solid line). The inset shows the NSLD depth profile of the film.

and $z = L$ (substrate–film interface), reducing the integral to $(1 + e^{iQL})$ on the right-hand side of equation (2.17). This has a periodicity of $2\pi/L$ and gives rise to the 'Kiessig oscillations' that are seen in figure 2.8(c). Figure 2.8(c) shows the specular unpolarized NR data (solid circles) and corresponding fit (line) for a single iron phthalocyanine (FePc) film grown on a glass substrate. The repetition of Kiessig oscillations is evident from figure 2.8(c) as a consequence of the interference of neutrons reflected from two interfaces (film–air and substrate–film). The momentum transfer difference (ΔQ) between the two consecutive Kiessig oscillations gives the total thickness of the film using the relation $L = 2\pi/\Delta Q$. The inset of figure 2.8(c) shows the NSLD depth profile of FePc film on a glass substrate.

A brief discussion of 'periodic bilayers' or multilayers is relevant at this point since such multilayers have been studied extensively for understanding various interface properties [13–22]. In a periodic bilayer, two components (or more) of a certain thickness are deposited alternately on a substrate. In the case of two components, A and B, the sequence is $ABABABAB$——. This periodic sequence can be compared to the periodicity seen in atomic arrangements in a crystal, with only a difference in the length scale of periodicity (in the case of a multilayer the bilayer thickness of ~ a few tens of nanometres

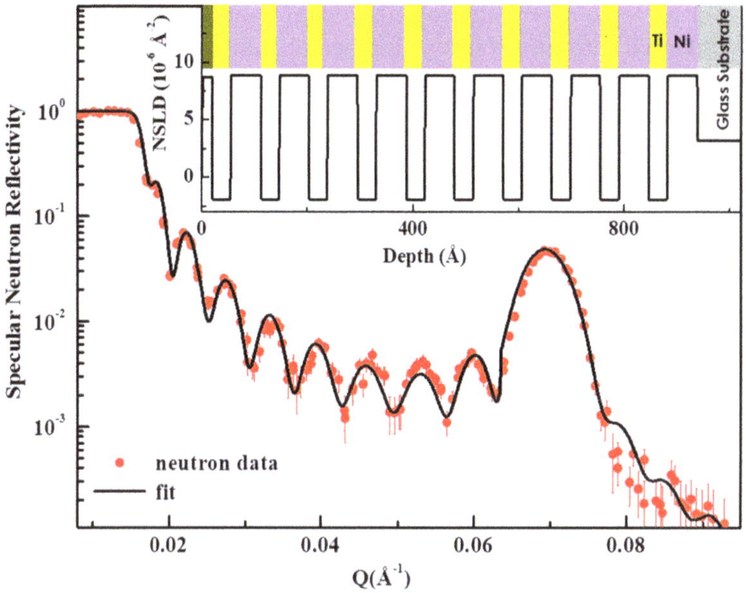

Figure 2.9. Specular unpolarized NR data from a Ni/Ti periodic multilayer with ten bilayers. The Bragg peak indicates the periodicity of the bilayer sequence. The inset depicts the NSLD depth profile of the Ni/Ti multilayer which best fits (solid line) the NR data.

is repeating, whereas in atomic arrangements there is interplanar spacing of ~ a few Å). The bilayer periodicity (repeatability) gives rise to Bragg peaks in a reflectivity experiment, but at much lower angles, since the length scales associated with thin film depositions lie in the nanometre range, inversely proportional to the Bragg angle. Figure 2.9 shows the experimental unpolarized NR data from a periodic Ni/Ti multilayer sample with ten bilayers. The Bragg peak is due to the periodicity of the layers. The position of the Bragg peak allows one to estimate the thickness of the bilayer (the equivalent of lattice spacing in a crystal). The Ni/Ti multilayer was grown on a glass substrate. The fit to NR data (solid line in figure 2.9) revealed the depth-dependent structure as shown in the inset of figure 2.9.

There are two approaches for generating reflectivity (both XRR and NR) patterns theoretically from a multilayer sample, which are named Parratt's formalism [11] and the matrix method [23]. Parratt's formalism [11] has been used extensively for both x-rays and neutrons, whereas the matrix method was used mostly for NR. Next, we will give a brief description of Parratt's formalism, which along with matrix method is highly used to generate theoretical reflectivity from neutrons [23].

2.3.2.1 Parratt's formalism: reflectivity profile for an N*-stratified homogeneous medium*

Parratt's formalism [11] allows one to calculate the reflected intensity from an idealized stratified medium of known layer thickness and density. This formalism was first introduced by Parratt [11] for x-ray reflectometry and it is used extensively for both neutrons and x-rays. It has the advantage of providing the correct expression for all regions of scattering since any density profile can be modelled by slicing the material into an arbitrary number of thin layers. A multilayer sample consisting of N layers, $i = 1, \ldots, N$, is shown in figure 2.10(a), where the refractive index, thickness, propagation angle, and interface roughness parameter of the ith layer are n_i, d_i, θ_i and σ_i, respectively. The Fresnel reflectivity between the $(i-1)$th and ith (ith lamina of the stratification) layer can be calculated from the continuity of the wavefunction ψ and its derivative $d\psi/dz$ at the interface between the two layers. The phase factor that is defined in the middle of the two surfaces of the ith medium is $a_i = e^{-iQ_i d_i/2}$, with d_i = the thickness of ith layer and $Q_i = \left(2\pi/\lambda\right)(n_i^2 - \cos^2\theta_i)^2$ is the z-component of the wavevector.

For each layer, the wavefunction consists of an incident amplitude $\psi_n(z)$ and a reflected amplitude $\psi_n^R(z)$ and one can write the continuity equation for the wavefunction and its derivative at the interface of the ith and $(i-1)$th layers:

$$a_{i-1}\psi_{i-1} + a_{i-1}^{-1}\psi_{i-1}^R = a_i^{-1}\psi_i + a_i\psi_i^R \tag{2.18a}$$

$$[a_{i-1}\psi_{i-1} - a_{i-1}^{-1}\psi_{i-1}]Q_{i-1} = \left[a_i^{-1}\psi_i - a_i\psi_i^R\right]Q_i. \tag{2.18b}$$

Solving (2.18a) and (2.18b) we obtain a recursion formula:

$$R_{i-1,\,i} = a_{i-1}^4 \left| \frac{R_{i,\,i+1} + F_{i-1,\,i}}{1 + R_{i,\,i+1}F_{i-1,\,i}} \right|, \tag{2.19}$$

where $R_{i,\,i+1} = a_i^2 \frac{\psi_i^R}{\psi_i}$ and $F_{i-1,\,i} = \frac{Q_{i-1} - Q_i}{Q_{i-1} + Q_i}$.

Using the above recursion relation one can obtain reflection amplitude $R_{1,2} = \frac{\psi_1^R}{\psi_1}$, at the air–film interface for a given multilayer. So the reflectivity of the system

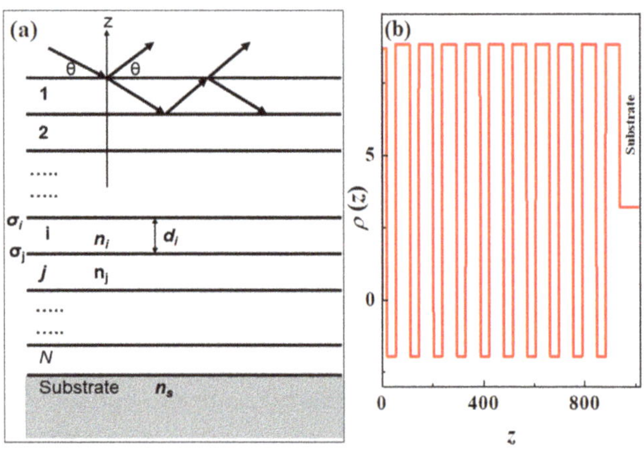

Figure 2.10. (a) A multilayer stack containing N layers, where the refractive index, thickness, propagation angle, and interface roughness parameter of the ith layer are n_i, d_i, θ_i, and σ_i, respectively. (b) Scattering length density, $\rho(z)$, of a multilayer is depicted as a function of z, as a histogram. Homogeneity of $\rho(z)$ is assumed in each layer.

(smooth interfaces) is given by $I = | R_{12} |^2$. The reflectivity for rough multilayer can be calculated by considering a static 'Debye–Waller' like factor (equation (2.16)) for reflectance at each interface in the multilayer and using the above recurrence relation (equation (2.19)). However, the recursion relation (equation (2.19)) has to be solved sequentially starting from the bottom (substrate layer) with the assumption that the reflectivity is zero at the bottom of the substrate and $R_{N,\,N+1} = 0$, where we have considered N number of layers for evaluating the reflectivity at the air–film interface. With this assumption, one can build the reflected amplitude upward, at every interface between the layers using equation (2.19). Parratt's formalism is the most-used technique for computing the reflectivity profile, for thin film heterostructures (single layer/multiple layers etc) by slicing the layers into lamina as shown in figure 2.10. The model density for XRR and NR is the parameters of the model that are varied to obtain best fits to the experimental data.

2.4 Specular polarized neutron reflectometry

Specular NR can be performed in either an unpolarized mode or polarized mode, designated as polarized neutron reflectometry (PNR). So far, we have restricted our discussion to unpolarized NR, where the neutron is only sensitive to the nuclear potential (equation (2.1)) of the medium, but the neutron is a spin ½ nuclear particle with a magnetic moment of $-1.91\ \mu_N$, where μ_N is a nuclear magneton. The magnetic moment of a neutron interacts with the magnetization in a sample. In the case of neutron diffraction (ND) experiments for unravelling magnetic lattice, the diffraction takes place from the crystallographic structure as well as from the magnetic long-range order present in the lattice. Using ND measurements, a neutron is considered to be a unique tool for extracting magnetic structures at an atomic length scale. In the case of reflectometry at a mesoscopic length scale, we need to consider the interaction of the neutron spin magnetic moment $\bar{\mu}$ with the local magnetic field in the sample given by $\bar{B}(r)$. The interaction potential is given by

$$V_{\text{magnetic}} = -\bar{\mu} \cdot \bar{B}(r). \tag{2.20}$$

With a magnetic sample, the potential energy can now be written as a summation of nuclear potential and a magnetic potential, given by

$$V_{\text{Total}} = V_{\text{nuclear}} \pm V_{\text{magnetic}} = V_{\text{nuclear}} \mp \bar{\mu} \cdot \bar{B}(r). \tag{2.21}$$

In general, the neutron magnetic moment vector can be represented as a combination of Pauli spin matrices and can be written as $\bar{\mu} = -1.91\mu_N\bar{\bar{\sigma}}$ with $\bar{\bar{\sigma}}$ is defined as a combination of Pauli spin matrices and $\mu_N = \frac{e\hbar}{2m_n C}$, nuclear magneton. In an actual experimental set up a magnetic field $\bar{H}$ is applied in the plane of the sample and its direction is assumed as the reference direction. The neutron moment is aligned either along (+) or opposite (−) to the applied field. The positive sign for the magnetic potential, which adds up to nuclear potential, is taken when the neutron moment is *parallel* to the field (spin-up) and vice versa. With the addition of the magnetic potential, the effective potential seen by neutrons in a magnetic medium is modified and is shown schematically in figure 2.11.

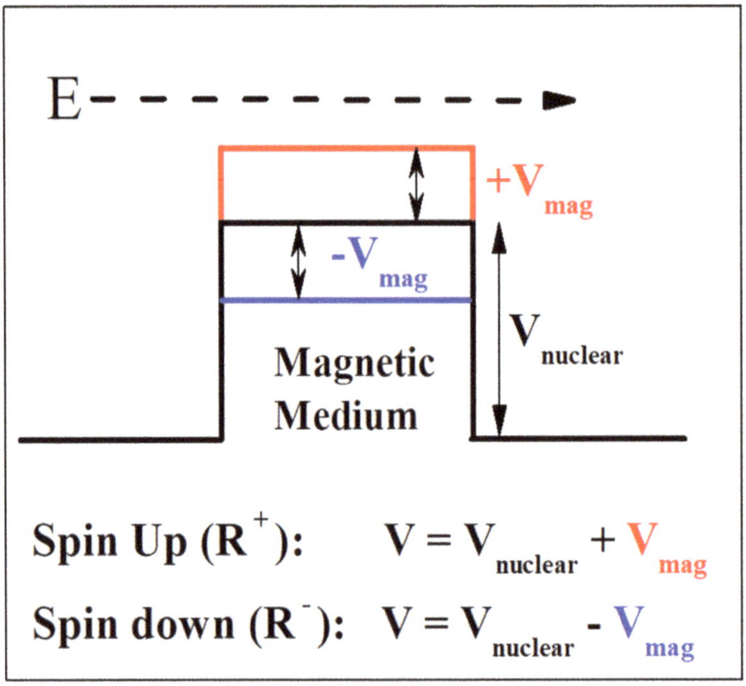

Figure 2.11. The potential seen by a neutron of energy E is a combination of the nuclear potential (V_{nuclear}) and the magnetic potential (V_{mag}) that gets added to or subtracted from the nuclear potential depending on the polarization of the incident neutron beam, with respect to the applied magnetic field on the sample.

Analogous to the nuclear scattering length b_{coh}, a magnetic scattering length b_M can be defined. This magnetic scattering length is related to μ_S, the magnetic moment per atom in a magnetic medium, expressed in units of Bohr magnetrons, according to

$$b_M = \frac{m_n \mu_n \mu_0}{2\pi\hbar^2} \mu_S. \tag{2.22}$$

So the magnitude of magnetic potential can now be written in terms of b_M, the magnetic scattering length:

$$V_m = \frac{2\pi\hbar^2}{m_n} N b_M, \tag{2.23}$$

where N is the number (nuclei) density and Nb_M is defined as magnetic scattering length density (MSLD). So, the total interaction potential for neutrons in a magnetic medium can be written in the form

$$V_{\pm} = V_n + V_m = \frac{2\pi\hbar^2}{m_n} N(b_{\text{coh}} \pm b_M) = \frac{2\pi\hbar^2}{m_n}(\rho_{\text{nuc}} \pm \rho_{\text{mag}}), \tag{2.24}$$

where the (+) and (−) signs correspond to the spin-up and spin-down neutrons with respect to sample magnetization.

The refractive index and critical angle for a neutron in the magnetic medium can be written as

$$n^{\pm} = 1 - \frac{\lambda^2}{2\pi}N(b_{\mathrm{coh}} \pm b_M); \theta_{\mathrm{c}}^{\pm} = \lambda\sqrt{\frac{N(b_{\mathrm{coh}} \pm \mathrm{b}_M)}{\pi}}. \tag{2.25}$$

In general, the reflection geometry with an applied magnetic field $\vec{H}$ is shown in figure 2.12 in a three-dimensional schematic. Similar to figure 2.4, the reflecting plane is the x–z plane and z is normal to the plane of the film. The applied magnetic field is along the y-direction and the polarization of the incident neutron will be considered positive (+) when it is along the applied field and minus (−) when it is opposite to the applied field. The local magnetic field, considered in magnetic potential seen by neutrons in a magnetic medium (in equation (2.20)) consists of the applied field, $\vec{H}$, and the magnetization of the sample, $\vec{M}$, given by (we have suppressed the explicit dependence of the parameters on coordinate r for convenience)

$$\vec{B} = \vec{H} + 4\pi\vec{M}. \tag{2.26}$$

The applied magnetic field $\vec{H}$ is continuous across the interface and does not add to any scattering length contrast in PNR. The magnetization of the sample $\vec{M}$ is the

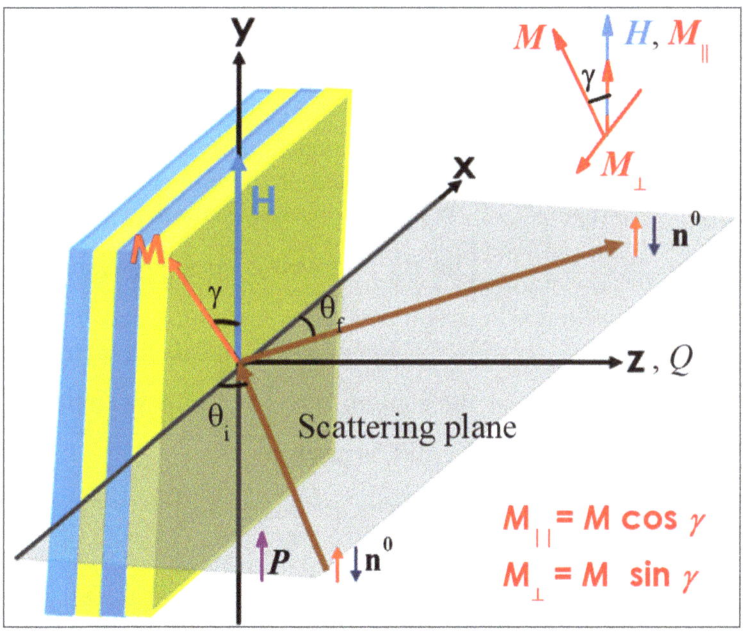

Figure 2.12. Reflection geometry for polarized neutron reflectivity experiment. The reflecting plane is the x–z plane here with the z-axis normal to the sample plane. The applied magnetic field H is along the y-axis and in the plane of the sample. Neutron polarization is defined as '+' when its moment is parallel to the field and '−' when it is antiparallel.

component of interest for PNR. Also, the magnetization $\overrightarrow{M}$ has been shown as in-plane in figure 2.12, since any out-of-plane component of the magnetization does not have any contribution to neutron scattering [24].

The general principle of a PNR experiment is illustrated in figure 2.12, in which a polarized neutron beam with polarization $\boldsymbol{P}$ collinear to the applied field (H) falls onto the magnetic thin film surface at an angle of incidence θ_i. The in-plane magnetization (M) of the film is at an angle γ with respect to the applied field. In the case of PNR, the reflectivity can be measured in two modes. (i) When the angle ($\gamma = 0$) of the rotation of magnetization is zero, this case arises when the applied field is sufficient to align the magnetization of the film along the applied field and this measurement will be termed as PNR without polarization analysis. We measure spin-up (R^+) and spin-down (R^-) reflectivities in this case and the PNR measurement provides depth-dependent magnetization information of the thin film heterostructures. (ii) In the other mode, the γ is finite suggesting that the magnetization vector has both magnitude and direction, which require spin analysis of the reflected neutron beam and it is termed as PNR with polarization analysis. We will discuss these two cases separately in the following section.

2.4.1 PNR without spin polarization analysis

In the case of PNR without polarization analysis, the in-plane magnetization, $\overrightarrow{M}$ is along the applied field $\overrightarrow{H}$ ($\gamma = 0$) and the neutron moment is either parallel or antiparallel to the applied field. Usually, the thin film samples used in PNR are soft magnets and an applied magnetic field in the range of ~2 kG is enough to align the sample's moment along the applied magnetic field. Under this condition, the magnetic potential seen by the neutron is due to the total moment $\overrightarrow{M}$ that either gets added to or subtracted from the nuclear potential (figure 2.11) depending on the neutron's polarization. There is no component of the moment normal to the field, since the angle $\gamma = 0$ (figure 2.12). We may write a two-component wavefunction by a column matrix $\psi(Z) = \begin{bmatrix} \psi_{+(Z)} \\ \psi_{-(Z)} \end{bmatrix}$ for the polarized neutron beam and construct the one-dimensional Schrödinger equation similar to equation (2.8a). The corresponding potential, in this case, may also be represented by a column matrix corresponding to the two polarizations of the neutron, $V = \begin{bmatrix} V_+ \\ V_- \end{bmatrix}$. The potential, in terms of nuclear and magnetic scattering lengths, is given by

$$V = \begin{bmatrix} V_+ \\ V_- \end{bmatrix} = \frac{2\pi\hbar^2}{m} N \begin{bmatrix} b_{\text{coh}} + b_{\text{M}} \\ b_{\text{coh}} - b_{\text{M}} \end{bmatrix} \tag{2.27}$$

When the magnetic moment in the sample is aligned along the applied field, we obtain two independent equations for $\psi_+(z)$ and $\psi_-(z)$. The equations are

$$\frac{d^2\psi_+(Z)}{dZ^2} + \frac{2m}{\hbar^2}[E - V_+(Z)]\psi_+(z) = 0 \tag{2.28a}$$

and

$$\frac{d^2\psi_-(Z)}{dZ^2} + \frac{2m}{\hbar^2}[E - V_-(Z)]\psi_-(Z) = 0. \tag{2.28b}$$

The solutions to the above equations give the propagation of the neutron beams with magnetic moments parallel (+) and antiparallel (−) to the applied field. We have dealt with the solution of such equations while deriving the reflectivity of an infinite medium in the case of an unpolarized neutron beam. In the present case, the solutions are different for the spin-up and spin-down neutrons due to the addition and subtraction of the magnetic potential with the nuclear potential in equations (2.28a) and (2.28b). The transmission and the reflection amplitudes are different for the two polarizations and can be denoted with the corresponding suffix for them while expressing the wavefunctions for spin-up and spin-down neutrons: $\psi_+(z) = e^{iqZ} + r_+e^{-iqZ}$ and $\psi_-(z) = e^{iqZ} + r_-e^{-iqZ}$. There are two sets of solutions for the two different neutron polarizations as a function of momentum transfer and also two different critical angles for the two polarizations as given in equation (2.25).

For calculating the specular PNR profile from a magnetic multilayer, there is formalism developed by Majkrzak [25] and Felcher *et al* [2] which describes the specific case of the neutron polarization axis being parallel to the sample surface. The Parratt [11] formalism and the matrix formalism given by Blundell and Bland [23] are also common to generate the theoretical PNR profile and both give identical profiles. Experimentally one measures two reflectivity profiles for the spin-up and spin-down neutrons as a function of wavevector transfer, Q_z, and to carry out such measurements a reflectometer should be equipped with different components, e.g. polarizer, flipper, etc [26]. A typical schematic for such an instrument is shown in figure 2.13 and different components are also marked. A supermirror polarizer provides a neutron beam polarization along the applied field direction. The spin flipper, which is an essential part of any polarized neutron instrument and rotates the neutron polarization by 180°, is used before the sample for changing the spin of the incoming neutrons.

Unpolarized neutrons from a neutron source (either reactor or spallation source) are polarized using a supermirror polarizer. The spin of the polarized (spin-up) neutron can be flipped (spin-down) by switching on the electric current (magnetic field) in the flipper. A polarized neutron falls on the sample at a grazing incidence

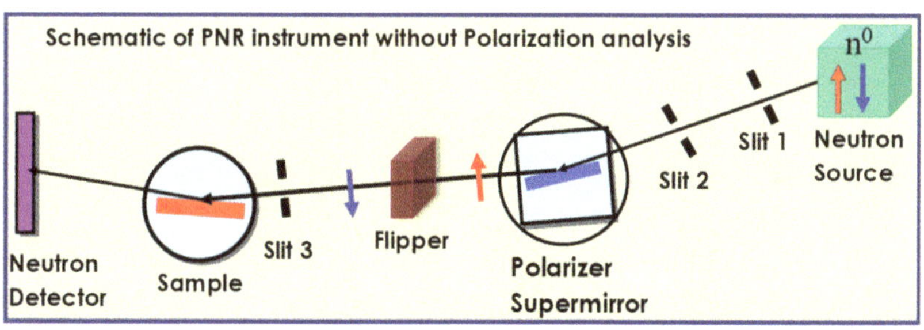

Figure 2.13. Schematic of a typical PNR instrument without polarization analysis, which measures the spin-dependent reflectivity (R^+: spin-up; R^-: spin-down).

and the reflected neutron is then detected on a neutron detector. Depending on the type of neutron source (a monochromatic neutron beam usually from a reactor source or neutrons with a spread of wavelength usually from a spallation source) one can cover the reflectivity as a function of momentum transfer by changing the angle of incidence(reactor)/wavelength (spallation).

Therefore, specular PNR without polarization analysis consists of two reflectivity patterns for two spins (R^+ and R^-) and analysis of data provides simultaneously the depth profile of both a nuclear component ρ_{nuc} and a magnetic component ρ_{mag} of scattering length density (SLD) [3, 4]. The SLD depth (along the film thickness, z-direction) profile $\rho_n(z)$ used to generate the PNR profile is [3, 4]

$$\rho_n(z) = \rho_{\text{nuc}}(z) \pm \rho_{\text{mag}}(z), \tag{2.29}$$

where

$$\rho_{\text{nuc}}(z) = \Sigma N_i(z) b_i; \ \rho_{\text{mag}}(z) = \Sigma N_i(z) p_i = C \Sigma N_i(z) \mu_i, \tag{2.30}$$

where the summation is over each type of atom in the system, $N(z)$ is the depth-dependent number density, b_{coh} is the nuclear scattering length, and μ is the magnetic moment of the scatterer atom in Bohr magnetons. The constant $C = 2.645\ \text{fm}/\mu_{\text{B}}$ or $2.9109 \times 10^{-9}\ \text{Å}^{-2}$ cc/emu [3].

Figure 2.14(a) shows the simulated PNR profiles for a Co/Cu multilayer, covering the first-order Bragg peak, using Parratt's formalism. Ten bilayers of Co and Cu with a thickness of 50 Å each (bilayer thickness = 100 Å) on a Si substrate are considered a multilayer for simulating the PNR profile. The schematic of the

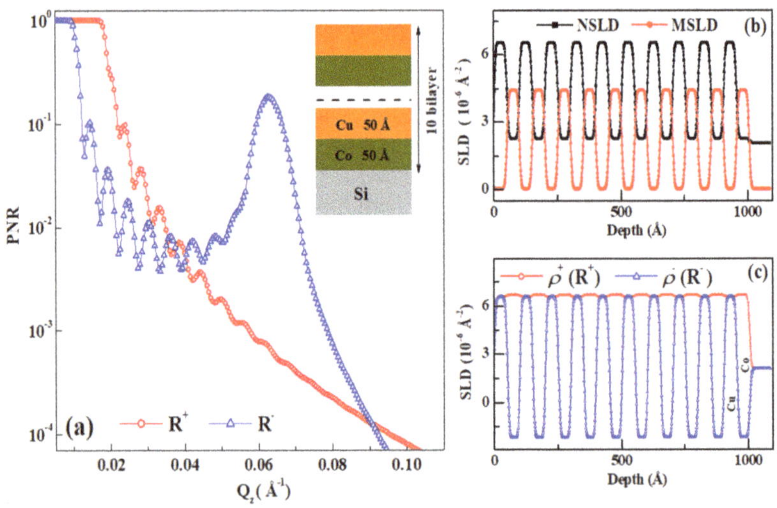

Figure 2.14. (a) Simulated PNR profiles without spin polarization analysis for a Si/[Co (50 Å)/Cu (50 Å)]$_{10}$ multilayer. The inset shows the schematic of the structure of the multilayer. The magnetization of each Co layer in the multilayer is aligned along the polarization of the neutron (applied field). (b) The NSLD and MSLD depth profile of the multilayer. (c) The sum and difference of NSLD and MSLD depth profiles responsible for spin-up and spin-down reflectivity profiles.

multilayer is shown in the inset of figure 2.14(a). Co moments are aligned along the applied in-plane field, thus $\gamma = 0$, in this case. Figure 2.14(b) shows the depth profile of NSLD and MSLD for the Co/Cu multilayer. Therefore by measuring spin-up (R^{+}) and spin-down (R^{-}) PNR one can estimate the depth profiling of structure (NSLD) and magnetization (MSLD). It is noted that the R^{+} profile does not show the Bragg peak corresponding to the periodicity of ten bilayers of Co and Cu. As mentioned earlier the R^{+} and R^{-} profiles depend on ρ^{+} (= NSLD + MSLD) and ρ^{-} (= NSLD – MSLD), respectively, shown in figure 2.14(c). The value of ρ^{+} shows minimum change (no contrast across interfaces) at interfaces as compared to ρ^{-} (figure 2.14(c)). The small contrast in SLD for spin-up neutrons (ρ^{+}) for Co/Cu multilayer (figure 2.14(c)) is the reason for the absence of the Bragg peak as well as the small intensity for Kiessig oscillation in the R^{+} (spin-up PNR) profile, whereas the SLD depth profile (ρ^{-}) responsible for the spin-down PNR (R^{-}) shows maximum change (large contrast) at the interfaces and thus shows a high-intensity Bragg peak as well as Kiessig oscillations (figure 2.14).

In addition, for an ideal multilayer (large contrast at the interfaces) there is a total $N_{\text{bl}} - 2$ (N_{bl} is the number of bilayers) Kiessig oscillation between two Bragg peaks. It is evident from figure 2.14(a) that a total of eight Kiessig oscillations between the critical angle to first-order Bragg peak in the R^{-} profile, suggesting the total number of bilayers is 10. Further, the Q (or Q_z) value corresponding to the first-order Bragg peak gives the thickness of the periodic bilayer using $t_{\text{bl}} = 2\pi/Q$. Similarly using this reciprocal relation one can obtain the total thickness of the film ($N_{\text{bl}}\, t_{\text{bl}}$) from the difference in Q between two consecutive Kiessig oscillations.

A comprehensive list of neutron reflectometer instruments available around the world is given in chapter 5. Figure 2.15 shows the image of the PNR instrument at Dhruva, BARC, India [26], which is designed for vertical sample geometry for PNR measurements without a polarization analysis option and is thus being used to investigate the magnetization depth profiles of thin films. This instrument at a reactor source (Dhruva) uses a monochromatic neutron beam with a wavelength of $\lambda \sim 2.9$ Å [4, 26], however, the wavelength of the neutron can be varied by rotating the monochromator. The reflectometer has been positioned on thermal neutron guide G2 in the Guide Tube Laboratory (GT Lab) of the Dhruva reactor and uses a linear position-sensitive detector (PSD), where only a sample is rotated to cover the desired Q range. Linear PSD also helps to collect both specular and off-specular reflectivity simultaneously at each angle of incidence.

2.4.2 PNR with spin polarization analysis

In the general case, when the orientation of the in-plane magnetization $\overrightarrow{M}$ is not along the applied magnetic field $\overrightarrow{H}$ (figure 2.12), one needs to consider the components (in-plane) of the magnetic moment parallel to the applied field $M_{\parallel} = \overrightarrow{M}\cos\gamma$ and the component normal to the field $M_{\perp} = \overrightarrow{M}\sin\gamma$, where γ is the angle between $\overrightarrow{M}$ and $\overrightarrow{H}$. The spin-dependent PNR [1, 3, 4] is defined by four reflectivities depending on the polarization channel, R^{++}, R^{--}, R^{+-}, and R^{-+}, where the first and second superscript denote the direction of the incoming and

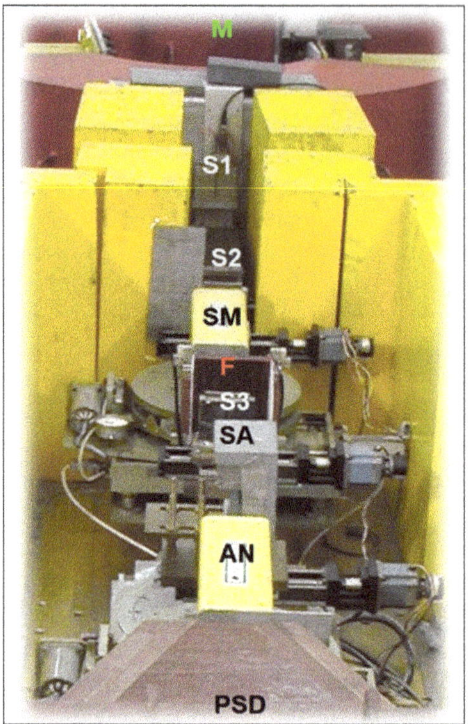

Figure 2.15. Image of the polarized neutron reflectometer at Dhruva, BARC, India. Different components of the instrument labelled in the image along with shielding are placed on a 5 m long table. This instrument is being used to characterise thin film magnetic samples using PNR without polarization analysis mode.

reflected neutron polarization as parallel (+) or antiparallel (−) with respect to the external field direction. The PNR with no change in the polarization upon reflection is termed as non-spin-flip (NSF) reflectivities (i.e. R^{++}, R^{--}). Whereas PNR with the change in the polarization of the reflected beam is termed as spin-flip (SF) reflectivities (i.e. R^{+-}, R^{-+}). It is evident from figure 2.12 that the NSF component of the reflected intensity has its origin in the parallel component, $M_{\parallel}$ of the sample magnetization. The normal component $M_{\perp}$ causes SF reflection due to the precision of the neutron moment around it, classically. One can re-build the direction and magnitude of the magnetic moment as a function of depth from NSF and SF components of PNR. The NSF component consists of R^{++} (up spin to up spin), R^{--} (down spin to down spin) and SF consists of R^{+-} (up to down) and R^{-+} (down to up). One can extract the values of $M_{\parallel}$ from the NSF data and $M_{\perp}$ from SF data in an experiment. The vector sum of the components allows us to obtain the direction and magnitude of the in-plane magnetic moment as a function of depth. PNR is possibly the only non-destructive technique that allows vector magnetometry at a mesoscopic length scale which will be discussed with examples later.

In this case, the potential will have components for NSF reflection given by the component of the magnetic moment of the sample along the applied magnetic field ($M_{\|}$) and components for SF reflection, given by the component of the magnetic field normal to the applied field ($M_{\perp}$) [1, 7, 8]. For the thin film stratification along the z-direction, the potential is dependent on z alone. We have introduced the Schrödinger equation for propagation of the neutron in equation (2.8a). As mentioned earlier, a reflection of a neutron is an elastic process and the energy of the neutron E is conserved during reflection. Also, the in-plane component of the wave vector, K_p, is conserved in a reflection. With the wavevector, K_z, normal to the plane of the film, the energy of the neutron is given by $E = -\frac{\hbar^2}{2m}\left[\overline{K}_Z^2 + K_P^2\right]$. We can write a Schrödinger equation in the z-component alone for the propagation of the neutron wave. In the case of specular reflection, the wavevector $K_Z = Q_z/2$, where Q_Z is the wavevector transfer in the reflection. The in-plane component of the wave vector remains unchanged and one can represent the potential for both NSF and SF reflection as a 2×2 matrix, $\begin{bmatrix} V_{++} & V_{+-} \\ V_{-+} & V_{--} \end{bmatrix}$, with the usual symbols for up (+) and down (−) neutrons [8]. Now we can write a combined Schrödinger equation in a matrix form for the two-component wavefunction $\psi(Z) = \begin{bmatrix} \psi_{+(Z)} \\ \psi_{-(Z)} \end{bmatrix}$ using the above potential with non-spin-flip (NSF) and spin-flip (SF) components for the propagation of neutron wave in the z-direction:

$$\frac{\partial^2}{\partial Z^2}\begin{bmatrix} \psi_+(z) \\ \psi_-(z) \end{bmatrix} + \frac{\hbar^2 Q_z^2}{4}\begin{bmatrix} \psi_+(z) \\ \psi_-(z) \end{bmatrix} - \frac{2m}{\hbar^2}\begin{bmatrix} V_{++}(z) & V_{+-}(z) \\ V_{-+}(z) & V_{--}(z) \end{bmatrix}\begin{bmatrix} \psi_+(z) \\ \psi_-(z) \end{bmatrix} = 0, \tag{2.31}$$

where Q_z is the z-component of momentum transfer in a specular reflection. Solutions to the master equation (2.31) encompass all possible reflections from a magnetized sample. The above matrix equation gives two coupled equations for ψ_+ and ψ_-, which are

$$\left[\frac{\partial^2}{\partial Z^2} + \frac{\hbar^2 Q_z^2}{4} - \frac{2m}{\hbar^2}V_{++}\right]\psi_+(z) - \frac{2m}{\hbar^2}V_{+-}\psi_-(z) = 0 \tag{2.32a}$$

and

$$\left[\frac{\partial^2}{\partial Z^2} + \frac{\hbar^2 Q_z^2}{4} - \frac{2m}{\hbar^2}V_{--}\right]\psi_-(z) - \frac{2m}{\hbar^2}V_{-+}\psi_+(z) = 0. \tag{2.32b}$$

If we put the SF potentials, $V_{+-}(z)$ and $V_{-+}(z)$, to zero, we obtain the previous equations (2.28a) and (2.28b) for the NSF case, discussed earlier.

The expression for the four components of the potential can be written in terms of the coherent NSLD, $\rho = Nb_{\mathrm{coh}}$, and components of the magnetic moment along the in-plane applied field and normal (in-plane) to it, written in terms of the MSLD components: $V_{++} = \frac{2\pi\hbar^2}{m}N\left[b_{\mathrm{coh}} + b_y\right] = \frac{2\pi\hbar^2}{m}N[b_{\mathrm{coh}} + b_{\mathrm{M}}\cos\gamma]$, here we have used a component of the magnetic scattering length along the magnetic field since the component of the magnetic moment $\overrightarrow{M}$, a vector quantity, is proportional to this. The other components of the potential are $V_{--} = \frac{2\pi\hbar^2}{m}\rho[b_{\text{coh}} - b_M \cos\gamma]$, $V_{\pm\mp} = \frac{2\pi\hbar^2}{m}\rho b_M \sin\gamma$. The NSF potentials have both nuclear and magnetic components and one tends to obtain both nuclear and magnetic moment densities from NSF data. The spin-flip potential $V_{\pm\mp}$ is of sole magnetic origin and from SF data one can extract the component of the magnetic field normal (in-plane) to the applied field. From the magnetization value obtained in SF reflectometry, one may vectorially add the components along and normal to the applied field $\overrightarrow{H}$ to obtain the direction and magnitude of the magnetic field in the magnetic thin film under study.

If we set the constants, $\hbar = m = 1$ and consider plane wave-like solutions, e^{iKz} for propagation in the z-direction, we obtain two related equations for ψ_+ and ψ_- with the following definitions for a set of wavevectors: $K_N^2 = 4\pi N b_{\text{coh}}$; $K_y^2 = 4\pi N b_M \cos\gamma$; $K_x^2 = 4\pi N b_M \sin\gamma$:

$$\left[\left(K^2 - K_z^2\right) + \left(K_N^2 + K_y^2\right)\right]\psi_+(z) + K_x^2\psi_-(z) = 0 \tag{2.33a}$$

$$K_x^2\psi_+(z) + \left[\left(K^2 - K_z^2\right) + \left(K_N^2 - K_y^2\right)\right]\psi_-(z) = 0. \tag{2.33b}$$

For a non-trivial solution, the determinant of the coefficients of ψ_+ and ψ_- should be zero and that gives us the possible solutions of the propagation vector K for a given Q_z. It is an algebraic equation of fourth order with the following solutions:

$$K^2 = K_z^2 - K_N^2 \pm \sqrt{K_x^4 + K_y^4}. \tag{2.34}$$

The term $\sqrt{K_x^4 + K_x^4} = 4\pi N b_M = K_M^2$ is a magnetic wavevector, by definition. The solutions are

$$K = \pm\sqrt{K_z^2 - K_N^2 \pm K_M^2}. \tag{2.35}$$

There are four solutions for the wavevector K in the medium for a given momentum transfer Q_z in a vacuum. Zabel *et al* [8] discussed the solutions to the equations for ψ_+ and ψ_- in detail. The negative solution for K indicates propagation in the backward direction from the medium which we will not discuss here. The two positive solutions, $K_\pm = \sqrt{K_z^2 - K_N^2 \pm K_M^2}$, indicate a neutron travelling in the medium from the vacuum (or air). Two solutions for two spin values of the incident neutron beam with respect to the applied magnetic field $\overrightarrow{H}$ are indicated in the above expression by the $\pm$ sign. We may define $Q_\pm = 2K_\pm$ to write expressions in terms of momentum transfers. The two neutron beams with opposite spins undergo dispersion in the magnetic media and the refractive index of the medium is different for the two spins. It needs to be mentioned that equations (2.33a) and (2.33b) have dependence on the angle γ that the magnetization vector $\overrightarrow{M}$ makes with the applied magnetic field $\overrightarrow{H}$ or the direction of the neutron magnetic moment, but the final solution for the propagation vector (equation (2.35)) does not have any dependence on γ and depends only on the orientation of the neutron moment with respect to the applied field $\overrightarrow{H}$ (parallel (+) or antiparallel (−)). We may express two critical angles $\theta_{C\pm}$ for the two spins (±):

$$Q_c = Q \sin \theta_{C\pm} = \sqrt{4\pi N(b_{coh} \pm b_M)}. \tag{2.36}$$

The independence of the critical angle in relation to the angle between the magnetization $\overrightarrow{M}$ and the neutron spin was demonstrated experimentally by measuring the reflectivity data for various angles γ [27]. The direction of $\overrightarrow{M}$ was rotated in-plane by application of a field before measurement. It was found that the critical angles defined above remain unchanged for all the orientations. Solving equation (2.33) for a polarization-dependent wave function the amplitude of reflected and transmitted waves can be obtained by using boundary conditions and a matrix method [2, 3, 8, 25], which gives a reflectance matrix [8]

$$\begin{pmatrix} r^{++} & r^{+-} \\ r^{-+} & r^{--} \end{pmatrix} = \begin{pmatrix} R_f^+\cos^2\left(\frac{\gamma}{2}\right) + R_f^-\sin^2\left(\frac{\gamma}{2}\right) & (R_f^+ + R_f^-)\cos\left(\frac{\gamma}{2}\right)\sin\left(\frac{\gamma}{2}\right) \\ (R_f^+ - R_f^-)\cos\left(\frac{\gamma}{2}\right)\sin\left(\frac{\gamma}{2}\right) & R_f^+\sin^2\left(\frac{\gamma}{2}\right) + R_f^-\cos^2\left(\frac{\gamma}{2}\right) \end{pmatrix} \tag{2.37}$$

where the $R_f^\pm$ are the Fresnel reflection amplitudes for neutron polarization parallel (+) or antiparallel (−) to the magnetization of the film,

$$R_f^\pm = \frac{\frac{Q_z}{2} - \sqrt{\frac{Q_z^2}{4} - Q_{c\pm}^2}}{\frac{Q_z}{2} + \sqrt{\frac{Q_z^2}{4} - Q_{c\pm}^2}}. \tag{2.38}$$

The resulting spin-dependent reflectivities for a single magnetic film on a substrate are given as

$$R^{++} = |\, r^{++} |^2 = \frac{1}{4} |\, R_f^+(1 + \cos\gamma) + R_f^-(1 - \cos\gamma)|^2$$

$$R^{--} = |\, r^{--} |^2 = \frac{1}{4} |\, R_f^+(1 - \cos\gamma) + R_f^-(1 + \cos\gamma)|^2, \tag{2.39}$$

and

$$R^{+-} = R^{-+} = |\, r^{\pm\mp} |^2 = \frac{1}{4} |\, R_f^+ - R_f^- |^2 \sin^2\gamma.$$

The spin-dependent reflectivities from a multilayer can be calculated by recursive application of the transfer matrix and this formalism has been reported in several articles [2, 3, 8, 25].

For spin-dependent PNR with polarization analysis measurements, a reflectometer should be equipped with different components and a typical schematic for such an instrument is shown in figure 2.16. A supermirror polarizer provides a neutron beam polarization along the applied field direction. The two spin flippers, one before the sample and the other after the sample, are used to change the spin of the

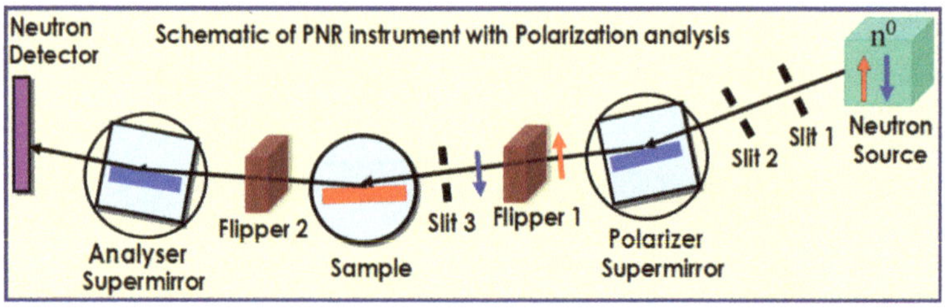

Figure 2.16. Schematic of a typical PNR instrument with polarization analysis, which measures the spin-dependent reflectivities, i.e. NSF (R^{++} and R^{+-}) and SF (R^{+-} and R^{-+}). Different components are marked in the schematic.

Table 2.1. Neutron spin states for different flipper settings for a reflectometer with a polarization analysis option.

Flipper 1	Flipper 2	Measured reflectivity
Off	Off	R^{++}
Off	On	R^{+-}
On	On	R^{--}
On	Off	R^{-+}

incoming and reflected neutrons. Typically a supermirror analyser of similar specification as that of a polarizer is used, to pass neutrons of a particular spin to the detector. The spin-dependent reflectivity can be selected by using different flipper settings as given in table 2.1, the flipper 'on' means applying a magnetic field perpendicular to the spin of the neutron for flipping its polarization by 180° (i.e. spin-up to spin-down or vice versa).

Figure 2.17 shows the simulated spin-dependent PNR profiles with spin polarization analysis for the same Co/Cu multilayer discussed in the previous section. Ten bilayers of Co/Cu multilayer of a bilayer thickness of 100 Å (schematic of the multilayer structure is shown in the inset of figure 2.14(a)) and the magnetic moment of each Co layer is at an angle of 30° (i.e. $\gamma = 30°$) with respect to an applied in-plane magnetic field (or polarization of neutron) is considered here. In figure 2.14 we assumed the Co moments are aligned along the field direction (i.e. $\gamma = 0°$). Since we have definite magnetization ($\sim M\sin\gamma$) along the normal (x-axis in figure 2.17) to neutron polarization (or applied field), we have observed SF PNR signals ($R^{\pm\mp}$) in addition to the NSF PNR signal ($R^{\pm\pm}$). Thus PNR data with polarization analysis can be used to investigate the depth-dependent magnetic structure (magnetization vector of each magnetic layer) from a multilayer.

2.5 Coherence area and resolution in neutron reflectometry experiments

The coherence of a radiation source (probe) is characterized by its ability to produce interference contrast in space (spatial coherence) and time (temporal coherence).

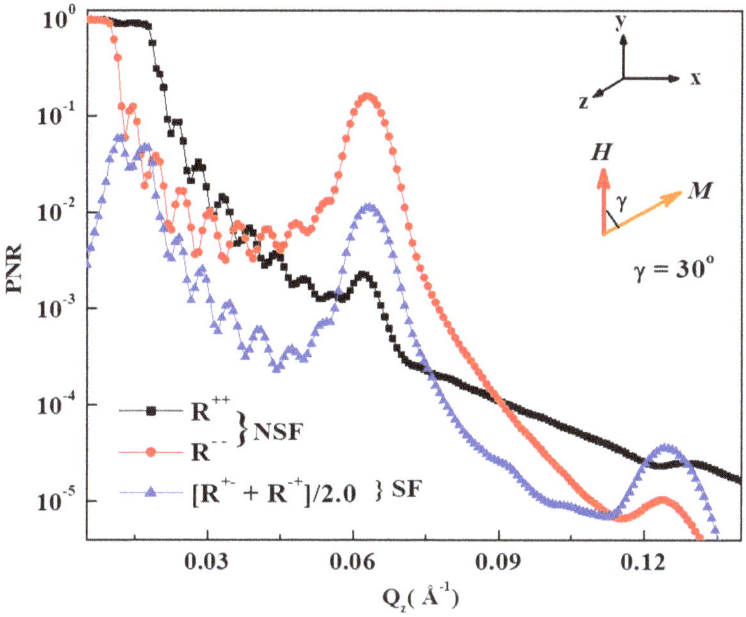

Figure 2.17. Simulated spin-dependent reflectivities from a Si/[Co (50 Å)/Cu (50 Å)]$_{10}$ multilayer assuming the angle between the Co magnetization and applied field is 30° in each Co layer.

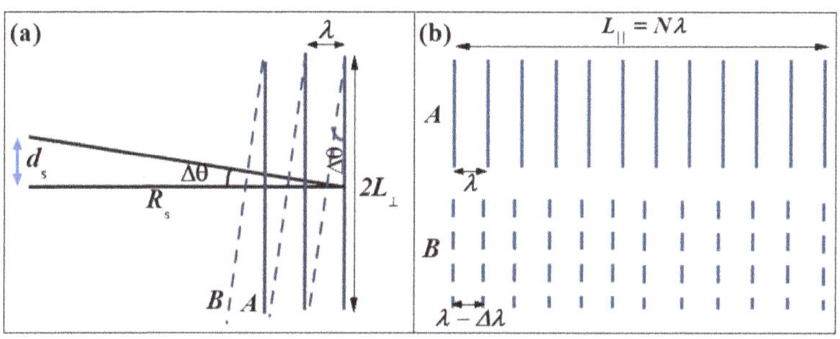

Figure 2.18. (a) Two wavefronts A and B are emitted from the two ends of a radiation source of size d_s and observed at a distance R_s. At one end of a detector of width $2L_\perp$ the waves are in phase and at the other end they are phase-delayed by exactly λ defining the transverse coherence length $L_\perp$. (b) A wave A of wavelength λ and a wave B of a shorter wavelength $\lambda - \Delta\lambda$ have the same amplitude at a distance $L_{||} = N\lambda$, called the longitudinal coherence length.

While temporal fluctuations are too rapid to be detected, the spatial coherence of a source can be quantified by defining appropriate coherence lengths [28] and spatial coherence can be quantified by two values called transverse coherence length ($L_\perp$) and longitudinal coherence length ($L_{||}$). Figures 2.18(a) and (b) show the cases of transverse and longitudinal coherence, respectively, for two wavefronts A and B. The transverse coherence length ($L_\perp$) is defined as the distance in the observation plane at which two wavefronts A and B, which are emitted from the two ends of a radiation source of size d_s and observed at a distance R_s, are phase-delayed by exactly the wavelength λ. It is evident from figure 2.18(a) that the angle $\Delta\theta$ is subtended by the radiation source at the observation plane and can be defined as $\tan\Delta\theta = \frac{\lambda}{2L_{\perp}} = \frac{d_S}{R_S}$, suggesting a transverse coherence length $(L_{\perp}) = \frac{\lambda R_S}{2d_S}$. In the case of longitudinal coherence length $(L_{\parallel})$, as illustrated in figure 2.18(b), it is the shortest distance between two positions at which a wavefront A of wavelength λ and a second wavefront B of a shorter wavelength $\lambda - \Delta\lambda$ have equal amplitudes, which is possible if $N + 1 = \frac{\lambda}{\Delta\lambda}$ (where N is the number of wavefronts). The longitudinal coherence length is then defined as $L_{\parallel} = N\lambda = \frac{\lambda^2}{\Delta\lambda}$. To improve coherence length for an x-ray source a pinhole beam size with a diameter of a few hundred microns and an energy resolution $(\Delta E/E)$ of ~10^{-4} can be used. Using a pinhole or cross slit of diameter $d_s = 0.1$ mm as an x-ray source of energy 8 keV and energy resolution $(\Delta E/E)$ of ~10^{-4} at 1 m from the detector and the transverse and longitudinal coherence lengths are 775 and 1550 nm, respectively.

Most of the modern neutron reflectometers, either steady state (reactor source) or time-of-flight (spallation), are equipped with a well-collimated neutron beam using a rectangular narrow entrance slit for the incident beam and with a PSD measuring the reflected beams. Neutron beams are intrinsically incoherent. Collimation and monochromatization, i.e. angular and wavelength resolution (uncertainty), ensure that neutrons with a close phase relationship make the neutron coherence volume/area. As shown in figure 2.19, the reflectometers are designed with a scattering plane as a horizontal z–x plane and the PSD covers the z–y plane with the glancing angle of incidence (θ_i) and reflection (θ_f). The short and long (and wide open) sides of the rectangular slit (see figure 2.19) are oriented along the z- and y-directions, respectively. The slit, parallel to the y-direction, enhances the intensity and this dimension is also used to enhance the intensity by focusing the incident beam onto

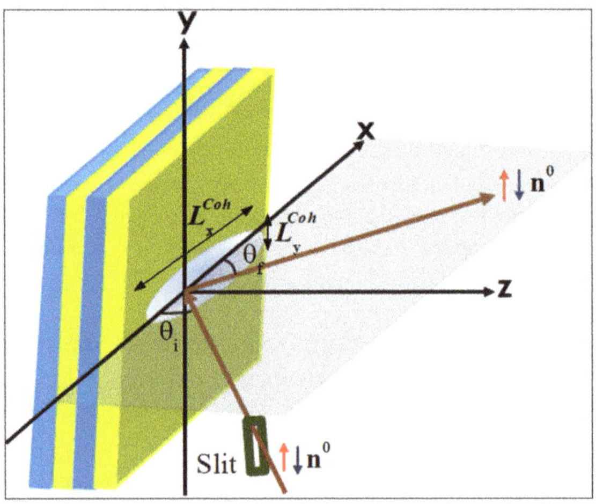

Figure 2.19. Schematic of the sample and beam geometry for PNR experiments with polarization analysis along the y-axis. The elongated ellipsoid indicates the coherence volume of neutrons defined by the beam divergence and the wavelength spread.

the sample surface. A combination of slits placed at a distance decides the angular divergence of the beam. Thus the angular resolutions (uncertainties) are drastically different in both directions: $\Delta\theta_x = \Delta\theta_i + \Delta\theta_f \ll \Delta\theta_y$. A wavelength ($\lambda$) selection using a monochromator makes the beam coherent along its propagation direction over a length $L_{\|} = \frac{\lambda^2}{2\pi\Delta\lambda}\Theta$, where $\Delta\lambda$ is the uncertainty in the wavelength and Θ is the phase difference of two coherent wavefronts. Along the direction perpendicular to its propagation direction, coherence length can be obtained using $L_{\perp} = \frac{\lambda}{2\pi\Delta\theta_i}\Theta$. The coherence ellipsoid (shown in figure 2.19) is the volume in real space that contributes to constructive/destructive inference.

The reflected intensities from different coherence volumes are added up incoherently to yield the total reflected intensity from a sample surface. Neutron coherence lengths or coherence volume or coherence area can be estimated from the uncertainties in the reciprocal space with the help of instrument resolutions. The projection of coherence volume onto the sample surface is estimated as a coherence area $A_{\rm coh}$ ($= L_x^{\rm coh} L_y^{\rm coh}$, where $L_x^{\rm coh}$ and $L_y^{\rm coh}$ are the coherence lengths along x- and y-directions, respectively) using angular divergences $\Delta\theta_i$, $\Delta\theta_f$ and $\Delta\theta_y$, which define the uncertainty in the momentum transfer ΔQ_x and ΔQ_y in the corresponding in-plane Q_x- and Q_y-directions. These uncertainties in momentum transfer within a small angle limit can be written as

$$\Delta Q_x = \frac{2\pi}{\lambda}\sqrt{(\theta_i\Delta\theta_i)^2 + (\theta_f\Delta\theta_f)^2}\,\tilde{}\,\frac{1}{L_x^{\rm coh}}; \quad \text{and} \quad \Delta Q_y = \frac{2\pi}{\lambda}\Delta\theta_y\,\tilde{}\,\frac{1}{L_y^{\rm coh}}. \tag{2.40}$$

In general, within PNR, scattering along the y-direction is not observed due to the relaxed collimation $\Delta\theta_y$, which is associated with large uncertainty in the limit $\Delta Q_y \gg \Delta Q_x$, and suggesting $L_y^{\rm coh} \ll L_x^{\rm coh}\left(=L_{\|}\right)$. Thus under PNR experimental conditions the coherence length $L_x^{\rm coh}$ ranges from a fraction of a millimetre to a few micrometres, whereas the coherence length, $L_y^{\rm coh}$, typically has an upper limit of 10 nm. The coherence length along the z-axis is determined by the uncertainty ΔQ_z in the wave vector transfer Q_z and it can be estimated by the uncertainties of the incident and exit angles using $\Delta Q_z = \frac{2\pi}{\lambda}\sqrt{(\Delta\theta_i)^2 + (\Delta\theta_f)^2}\,\tilde{}\,\frac{1}{L_z^{\rm coh}}$, which gives a coherence length ($=L_{\perp}$) of ~ a fraction of a micron for PNR experiments. The different coherence lengths in a typical PNR instrument thus suggest that $L_y^{\rm coh} \ll L_z^{\rm coh} \ll L_x^{\rm coh}$. In general the coherence area for the neutron beam is considerably smaller than the lateral dimensions of the thin film sample and the observed intensity is an incoherent summation of all the intensity coherently scattered from the coherence area distributed over the film. Therefore the large-scale lateral inhomogeneities of the sample are averaged and cannot be distinguished.

In reflectometry the reflected intensity is measured as a function of momentum transfer Q. The measured reflectivity is a convolution of reflectivity with the resolution of the instrument. To compare the measured data with the calculated data, first we have to correct it for the instrumental resolution. This can be done either by deconvoluting the measured data for experimental resolution or convoluting the theoretical profile with an experimental resolution before comparing it with experimental data. The instrumental resolution in small angle approximation is given by

$$\frac{dQ}{Q} = \sqrt{\left(\frac{d\theta}{\theta}\right)^2 + \left(\frac{d\lambda}{\lambda}\right)^2}. \tag{2.41}$$

The typical wavelength resolution ($\frac{d\lambda}{\lambda}$) for a single crystal monochromator-based reflectometer is ~1%–1.5%. The major contribution comes from the divergence of the neutron beam which can be varied by selecting different widths of slits and the typical Q resolution ($\frac{dQ}{Q}$) of a reflectometer can be chosen around 1%–3%. Usually, the reflectivity profile is measured for a constant value of $\frac{dQ}{Q}$ in the whole range of Q, which is done by choosing the appropriate combination of slits for various Q ranges.

2.6 Off-specular (or diffuse) neutron scattering

So far the discussion in this chapter has been limited to specular reflectometry, where the incident angle and the exit angle are equal and the momentum transfer vector is normal to the plane of the film. In addition, for specular reflectivity, either in unpolarized or in polarized mode, one obtains a physical or magnetic structure normal to the plane of the film as a function of depth in the heterostructure. Another class of experiments are possible, where the incident and the exit angles are unequal, as shown in figure 2.2, and are known as off-specular or diffuse neutron scattering (DNS). Off-specular (diffuse) x-ray scattering methods have been widely applied over many decades to characterize the morphology of rough surfaces and interfaces, particularly with the availability of sources of ever-increasing brilliance, x-ray synchrotron sources. Similar techniques using neutron beams have also become widespread, particularly for the study of magnetic multilayers. If we consider the reflecting plane as a z–x plane, as shown in figure 2.2, apart from the normal component of wave vector transfer, Q_z, there is also a component of wavevector transfer along the plane of the sample, Q_x. One uses this component to obtain the nature of interface morphology in terms of the in-plane height–height correlation function in such experiments as described below.

Let us assume that the film is homogeneous at the length scale being probed, except for the presence of a surface. The differential cross-section for scattering, $\frac{d\sigma}{d\Omega}$, of the neutron by a film having an NSLD of Nb with a rough surface as indicated schematically in figure 2.20, is given in the Born approximation [9]:

$$\frac{d\sigma}{d\Omega} = N^2 b^2 \int_V d^3r \int_V d^3r' e^{-iQ(\bar{r}-\bar{r}')}, \tag{2.42}$$

where Q is the momentum transfer in the scattering. The integration is over the volume of the sample. For simplification, the above volume integral can be reduced to a surface integral to deal with off-specular reflectivity [9]. One may consider a rough surface as a height fluctuation over a smooth surface with normal in the

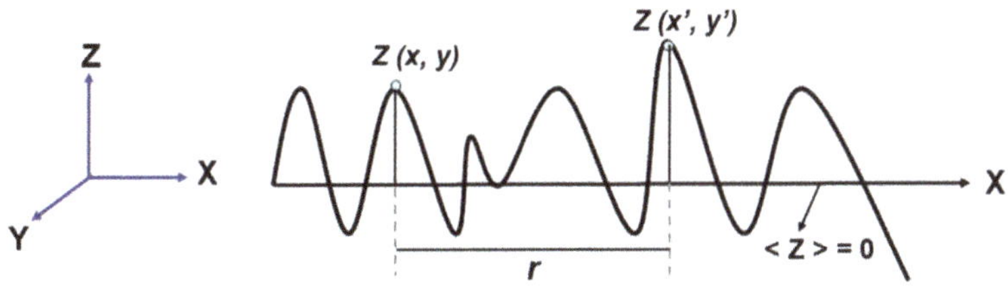

Figure 2.20. Variation of the height $z(x, y)$ of a rough surface with respect to an average smooth x–y plane, i.e. $\langle z\rangle = 0$, where z is normal to the surface.

z-direction (figure 2.20). The scattering cross-section from such a thin film sample can be represented as the following:

$$\frac{d\sigma}{d\Omega} = \frac{N^2b^2}{Q_Z^2}\iint dxdy \iint dx'\,dy'\, e^{-iQ_x(x-x')}e^{-iQ_y(y-y')}e^{-iQ_z[z(x,\,y)-z(x',\,y')]}. \qquad (2.43)$$

The integral is over the surface of the sample. The actual height of the rough surface with respect to the average smooth surface, given by, $z(x, y)$, depends on the local coordinate (x, y) on the surface. The dependence of the scattering cross-section in equation (2.43) on the height–height correlation is evident from the exponential terms in equation (2.43). The variation of the local height $z(x, y)$ with respect to the average smooth the x–y plane $[\langle z\rangle = 0]$ is shown in the schematic of figure 2.20.

The variation in the local height is responsible for the off-specular reflection of neutrons (and x-rays), causing deviation from Snell's law of reflection: $\theta_\mathrm{i} \neq \theta_\mathrm{f}$ (figure 2.2). An important assumption of the following treatment is that the height difference $[z(x, y) - z(x', y')]$ can be expressed as a Gaussian random variable $g(X, Y)$ [where $x - x' = X$ and $y - y' = Y$], which depends only on the relative coordinates $[X, Y]$. The $g(X, Y)$ is a height difference function [9] that can be expressed as

$$g(X,\ Y) = <[z(x,\ y) - z(x',\ y')]^2>. \qquad (2.44)$$

The angle bracket signifies the average over all the points in the plane separated by (X, Y). With $R = \sqrt{X^2 + Y^2}$, the distance between points, we may express

$$g(R) = 2\sigma^2\left[1 - e^{\left(-\frac{R}{\xi}\right)^{2h}}\right], \qquad (2.45)$$

where σ is the root-mean-square roughness of the surface that we seek to extract from off-specular reflectivity data and h is known as the Hurst parameter that defines the fractal dimension of the interface $D = 3 - h$ $(0 < h < 1)$. The above expression defines a correlation length ξ that sets two limits of the correlation function $g(R)$.

i. When $R \ll \xi$ we arrive at an expression $g(R) = CR^{2h}$, with C as a constant. This kind of correlation can be associated with *self-affine* rough surfaces, defined by Mandelbrot [29].
ii. In the other limit, when $R \gg \xi$, $g(R)$ saturates to the root-mean-square (rms) roughness (σ).

We may associate the correlation length, ξ, with the in-plane correlation length of the radiation (x-ray/neutron). This indicates that the average roughness that one may obtain from off-specular scattering will be an areal average over the plane of the sample indicative of the rms roughness defined above. With an expansion of the Q_z-dependent exponential in equation (2.43) we can express the diffuse scattering in terms of $g(X, Y)$:

$$\frac{d\sigma}{d\Omega} = \frac{N^2 b^2}{Q_z^2} L_x L_y \iint_S dX dY e^{-\frac{1}{2} Q_z^2 g(X,\, Y)} e^{-iq_x X} e^{-iq_y Y}. \tag{2.46}$$

The integrated intensity is over the area of $L_x L_y$ of the sample. It needs to be mentioned that the rms roughness that we obtain from the 'Debye–Waller' like factor (equation (2.16a)) from the fits of specular reflectivity data is a convolution of interface height fluctuation and interface alloy formation because both of these add to the *uncertainty* in the interface region. In the case of off-specular scattering, the rms roughness (σ) is dependent on only the height fluctuation at the interface and may be defined as a *true roughness* at the interface. This is additional merit of obtaining the roughness parameter, σ, from off-specular scattering. In this book we present instances where the roughness has a cut-off and the height difference saturates to a finite value $2\sigma^2$, the root-mean-square roughness. Sinha *et al* [9] have discussed different cases in detail where there is no cut-off and we refrain from discussing such cases here and suggest the above article to the interested reader.

In the Born approximation the reflection process is considered as scattering from independent scatterers being irradiated by an incident plane wave. In this approach, the differential cross-section is calculated for the scattering from the incident plane wave to the final plane wave. However, in the Born approximation, the prediction of the diffuse scattering behaviour breaks down for small values of the angle of incidence or at the total reflection region and one uses another approach called the distorted wave Born approximation (DWBA), which takes into account the process of refraction and multiple scattering [9]. The essence of the DWBA consists of an assumption that the initial state corresponds to a stationary wave field due to reflection from an ideal surface. The surface roughness acts as a perturbation and it causes the scattering from the initial to the time-reversed final state [30]. The detail of the DWBA approach for diffuse scattering is provided in different references [9, 31] and one can see these papers for a basic understanding of this approach.

Holy and co-workers [31] proposed the general DWBA formalism for the calculation of the intensity of the x-ray diffuse scattering from a multilayer thin film system. We have used this formalism for analysing the off-specular x-ray and neutron scattering data. This formalism allows one to evaluate the diffuse scattering profile in terms of a layer-specific height–height correlation function:

$$C_j(R) = \sigma_j^2 e^{-\left[\frac{R}{\xi_j}\right]^{2h_j}}, \tag{2.47}$$

where $C_j(R)$ is a layer-specific correlation function for the jth layer. The other parameters for the morphology of the jth layer are the correlation length (ξ_j) and

Hurst parameter (h_j). The height difference correlation function, $g(R)$, and correlation function, $C(R)$, are related by

$$g_j(R) = 2\sigma_j^2 - 2C_j(R). \tag{2.48}$$

Using these correlation functions one can obtain the differential cross-section for scattering, $\frac{d\sigma}{d\Omega}$, either by the Born approximation [9] or the DWBA [31]. Thus the off-specular intensity can be obtained using the following equation:

$$I_{\text{diff}} = I_0 \frac{\Delta\Omega}{A} \frac{d\sigma}{d\Omega}, \tag{2.49}$$

where A is the area of the beam, I_0 is the intensity of the incident beam, and Ω is the solid angle the detector subtends at the sample.

Here we will discuss off-specular unpolarized neutron scattering measurements from a Ni film for the estimation of height–height correlation (morphology) at different interfaces. Further, the surface morphology measured using diffuse neutron scattering will be compared with the atomic force microscopy measurements. The Ni film was a piece of a Ni guide in Dhruva, BARC, India [32]. We carried out a grazing incidence x-ray diffraction (GIXRD) for obtaining the crystallographic structure, which suggested the growth of polycrystalline Ni and no signature of an oxide layer. The GIXRD data are shown in inset (ii) of figure 2.21(a), while the

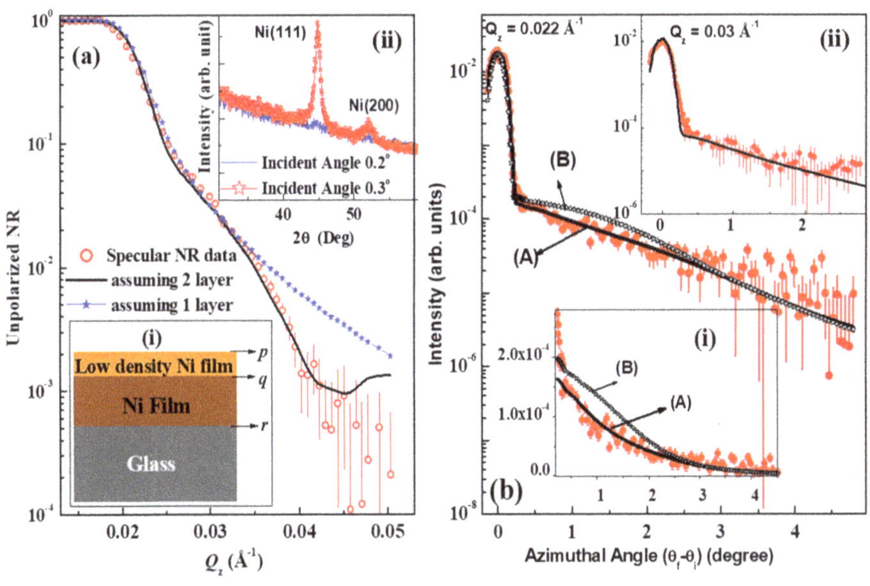

Figure 2.21. (a) Specular unpolarized NR data (open circles) and corresponding fits assuming two models, i.e. a single Ni layer and two Ni layers. The layer structure predicted by the best fit to NR data is shown schematically in the inset (i). The inset (ii) shows the GIXRD data at two angles of the incident of x-ray. (b) Off-specular unpolarized neutron scattering data (closed circles) at wavevector transfers 0.022 Å^{-1}. The magnified data at a limited angular range are shown in the inset (i). Similar data at a Q_z = 0.03 Å^{-1} is shown in the inset (ii). The solid lines (marked as (A)) indicate best fits assuming separate sets of roughness parameters, σ, ξ, and h (see the text) for the interfaces p and q. Attempting to fit a single set of parameters for both interfaces worsens the fit (marked as (B)). Reproduced with permission from [32]. Copyright 2006 Elsevier.

unpolarized specular NR measurements (figure 2.21(a)) were carried out to find the layer structure of the film [32]. Specular NR suggested that there is a low-density Ni layer of thickness ~235 Å at the air–film interface above a layer of near-bulk density (thickness ~1200 Å). GIXRD data confirmed that the low-density top layer is not an oxide layer and if there is an oxide layer then its thickness should be comparable to the roughness of the air–film interface (~9 Å, as obtained from specular NR).

The specular NR and off-specular neutron scattering experiments were carried out at the PNR instrument in Dhruva [26, 32]. As mentioned earlier this instrument uses a linear position-sensitive detector (PSD) to collect the reflected intensity at a certain reflection angle from a vertical thin film sample. Figure 2.21(a) shows the specular NR data with a fit assuming a single (blue star with line) and two (black solid line) Ni layer structures and it is evident from the data that the two-layer structure fitted the NR data best. Thus the results from GIXRD and specular NR data revealed a two-layer structure, shown in the inset (i) of figure 2.21(a), with three interfaces: (a) air–film interface, (b) low-density layer/high-density layer interface, and (c) film–substrate interface, which are depicted as p, q, and r in the inset (i) of figure 2.21(a).

The DNS data measured from the Ni film sample at two values of Q_z using the same PNR instrument are shown in figure 2.21(b). In this instrument the incident beam is vertical and the data are collected along the length of a PSD which may be considered the x-direction. This arrangement allows the measurement of DNS data at any angle of incidence. It needs to be mentioned that the top low-density layer and the high-density layer below were quite thick and we ignored any vertical correlation between the layer-specific parameters in our analysis. In this case, we attempted to fit three independent sets of layer-specific parameters (σ, ξ, and h) to fit the diffuse scattering data using the Holy formalism [31] for multilayers, which adopts the DWBA approach. The same and different morphological parameters for the two interfaces (p and q) in the inset (i) were considered in the attempt to fit the data and it is evident that different parameters for two Ni interfaces (interfaces p and q, inset (i), figure 2.21(a)) fitted the off-specular neutron scattering data best. The morphological parameters, σ, ξ, and h, for two interfaces, p (film–air interface) and q (film–low-density film interface), obtained from DNS data are 6.6 Å, 820 Å, and 0.52 (for the p interface) and 9.6 Å, 386 Å, and 0.89 (for the q interface), respectively.

The height distribution of the surface (film–air interface) was also measured using atomic force microscopy. Figure 2.22(a) shows the atomic force microscopy image of an area of 1.45 μm × 1.45 μm, over which the height measurement was done. We compared the parameters obtained from DNS for the air–film interface (surface) with that for the surface obtained from atomic force microscopy data. Figure 2.22(b) shows the height difference correlation function for the surface and corresponding fit (black solid line) using equation (2.48), which provided the surface morphological parameters of σ = 6.5 Å, ξ = 760 Å, and h = 0.48. The profile using the morphological parameters obtained from DNS is also plotted (red dashed line) in figure 2.22(b).

The excellent match between the parameters obtained from DNS and those obtained from atomic force microscopy in figure 2.22(b) indicates the usefulness of

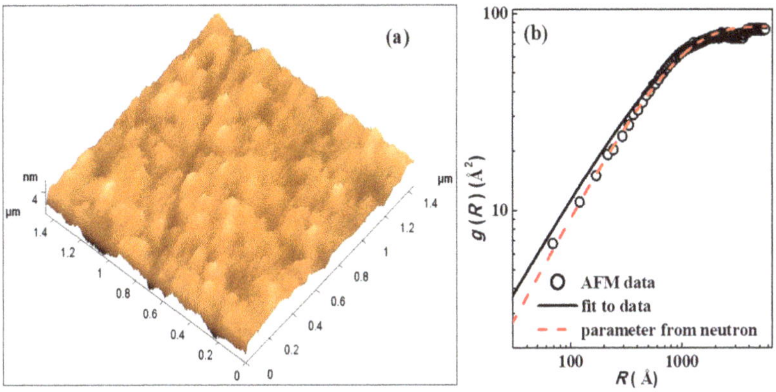

Figure 2.22. (a) Atomic force microscopy image of an area of 1.45 μ × 1.45 μ, over which the height measurement was carried out. (b) The height difference correlation function $g(R)$ (open circles) was obtained from the atomic force microscopy image. The black line is the fit to the atomic force microscopy data (see the text). The red dashed line is the $g(R)$ function using morphological parameters obtained from DNS for the air–film (surface) interface. Reproduced with permission from [32]. Copyright 2006 Elsevier.

DNS for obtaining interface morphological parameters. More importantly, one also obtains information regarding hidden interfaces that are not accessible by a technique such as atomic force microscopy. In the present case, the parameters for the interface (q), obtained from the DNS data suggest that this interface is much *smoother* with a roughness comparable to the surface. In addition, the study also suggested that the interface (p) shows a fractal nature with a dimension of 2.5, while the interface (q) represents a dimension of ~2.0. We have provided a few more examples of diffuse neutron and x-ray scattering in chapter 4, where these techniques have played an important role to investigate the interface morphology in thin film heterostructure systems of technological importance. In the next section we will describe the theoretical approach of polarized diffuse neutron scattering in BA and DWBA.

2.7 Polarized diffuse neutron scattering

2.7.1 Polarized DNS without spin polarization analysis under the Born approximation

In this section, we present a theoretical analysis of polarized DNS (PDNS) under BA using polarized neutrons without polarization analysis to separate the chemical and magnetic roughness morphology of a magnetic film. Using resonance off-specular x-ray scattering, Osgood *et al* [33] extended the treatment for the presence of magnetic roughness along with chemical roughness by using photons of two helicities with respect to sample magnetization, under the BA. We derived the expression for PDNS under the BA for neutrons of two polarizations [34] with respect to in-plane sample magnetization. Here we derive a general expression for PDNS under BA assuming the existence of a structural (chemical) and a separate magnetic boundary buried below it at each interface of the magnetic heterostructure. The schematic of a magnetic interface (boundary) buried below a physical interface and separated from it by a magnetic dead layer (Δ) is shown in figure 2.23.

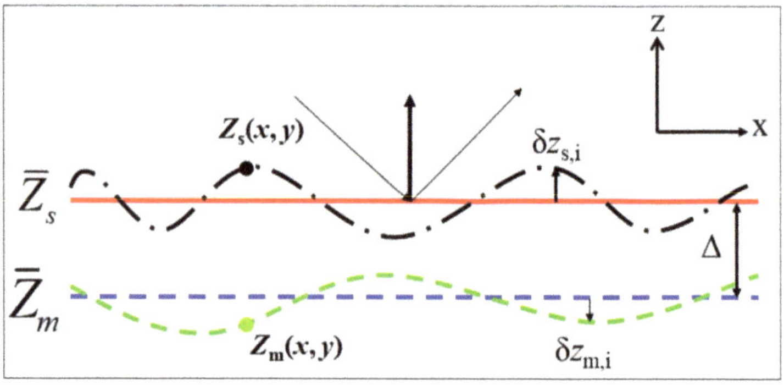

Figure 2.23. A schematic of the structural (chemical) and magnetic (virtual) boundaries at an interface. The height of the physical boundary (Z_S) and magnetic boundary (Z_m) with respect to the mean, smooth boundaries are shown and the separation (Δ) is a length scale for the magnetic dead layer at the interface. Reproduced with permission from [34]. Copyright 2009 IOP Publishing.

The actual boundaries (figure 2.23) can be then considered as really composed of two boundaries, a chemical boundary and a magnetic boundary, each with its average height, roughness, and correlation length. Here we discuss the presence of a 'magnetic roughness' together with physical roughness and its quantification using PDNS. The correlation term between the chemical and magnetic roughness for the chemical boundary and the magnetic boundary was defined similarly as defined by Osgood *et al* [33]. Using the potential seen by neutrons in a magnetic medium (equation (2.24)), the amplitude for grazing incidence neutron scattering is given by

$$A = \frac{Nb_N}{Q_z} \iint dxdy e^{-iQ_z Z_S(x,\, y)} e^{-iQ_x x} e^{-iQ_z y} \pm \frac{Nb_M}{Q_z} \iint dxdy e^{-iQ_z Z_M(x,\, y)} e^{-iQ_x x} e^{-iQ_y y}, \quad (2.50)$$

where $Z_s(x, y)$ and $Z_M(x, y)$ are the heights of the structural and magnetic interfaces, respectively (see figure 2.23). The Nb_N and Nb_M are the NSLD and MSLD, respectively.

The differential scattering cross-sections for the surface scattering can be written as [34]

$$\begin{aligned} \frac{d\sigma}{d\Omega} &= AA^* = \frac{N^2}{Q_z^2} \iint dxdy \iint dx'dy' \left[e^{-iQ_x(x-x')} e^{-iQ_y(y-y')} \right] \\ &\times \left[b_N^2 e^{-iQ_z\{Z_s(x,\, y) - Z_s(x',\, y_\prime)\}} + b_M^2 e^{-iQ_z\{Z_M(x,\, y) - Z_M(x',\, y')\}} \right. \\ &\left. \pm b_N b_M e^{-iQ_z\{Z_S(x,\, y) - Z_M(x',\, y')\}} \pm b_N b_M e^{-iQ_z\{Z_M(x,\, y) - Z_S(x',\, y')\}} \right]. \end{aligned} \quad (2.51)$$

The first two terms in equation (2.51) are purely structural and magnetic. The other terms are interference between structural and magnetic height functions and the $\pm$ indicates polarization of the neutron beam. Using the correlation function $C(X, Y)$ defined above and assuming the height fluctuations as Gaussian random variables, we can write equation (2.51) in the following compact form:

$$\frac{d\sigma}{d\Omega} = \frac{N^2 L_x L_y}{Q_z^2} [b_N^2 S_{ss} + b_M^2 S_{MM} \pm b_N b_M S_{SM}], \quad (2.52)$$

where L_x and L_y are lateral dimensions of the film in the (x, y) plane and the S_{ij} terms are the following:

$$S_{SS} = e^{-Q_z^2\sigma_S^2} \iint dXdY[e^{Q_z^2 C_{SS}(X,\ Y)} - 1]e^{-iQ_x X}e^{-iq_y Y} \tag{2.53a}$$

$$S_{MM} = e^{-Q_z^2\sigma_M^2} \iint dXdY[e^{Q_z^2 C_{MM}(X,\ Y)} - 1]e^{-iQ_x X}e^{-iQ_y Y} \tag{2.53b}$$

$$S_{SM} = e^{-Q_z^2\sigma_{SM}^2}e^{-iQ_z(\overline{Z}_s - \overline{Z}_M)} \iint dXdY[e^{Q_z^2 C_{SM}} - 1]e^{-iQ_x X}e^{-iQ_y Y}, \tag{2.53c}$$

where $\overline{Z}_S - \overline{Z}_M$ is the difference between the mean structural and magnetic interface heights, σ_S and σ_M are the root-mean-square Gaussian fluctuations of the structural and magnetic heights, quantifying the structural and magnetic roughness at the interfaces. The vector (X, Y) is the difference in coordinates between points $[X = x - x',\ Y = y - y']$. The correlation functions in equations (2.53a)–(2.53c), C_{SS}, C_{MM}, C_{SM}, are the structural (SS), magnetic (MM), and structure–magnetic (MS) functions, respectively. With $R = \sqrt{X^2 + Y^2}$, the scalar distance between points in (x, y), the correlation functions are defined as

$$C_{SS}(R) = < \delta Z_S(0)\delta Z_S(R) >;\ C_{MM}(R) = < \delta Z_M(0)\delta Z_M(R) >;$$

$$C_{SM}(R) = < \delta Z_S(0)\delta Z_M(R) >. \tag{2.54}$$

If we assume that both the structural and magnetic interfaces can be represented as self-affine fractals, the above correlation functions can be expressed as

$$C_{SS}(R) = \sigma_S^2 e^{-\left(\frac{R}{\xi_S}\right)^{2h_S}} \quad C_{MM}(R) = \sigma_M^2 e^{-\left(\frac{R}{\xi_M}\right)^{2h_M}}, \tag{2.55}$$

with the subscripts S and M denoting parameters related to the structural and magnetic interfaces and other parameters are already defined in equation (2.47), with the difference that now they relate to structural and magnetic interfaces. It also needs to be pointed out that in the above expression, equation (2.53), we have introduced a correlation function C_{SM} between the magnetic and structural interfaces. In the first approximation one can define the average magnetic and structure roughness for σ_{SM} as $\sigma_{SM}^2 = \frac{\sigma_S^2 + \sigma_M^2}{2}$. It may further be assumed that the structure–magnetic correlation length can be approximated as $\xi_{SM}^2 = \frac{\xi_S^2 + \xi_M^2}{2}$. It is evident from equation (2.53) that if there is no correlation between the structural and the magnetic interfaces, then $C_{SM} = 0$ and $S_{SM} = 0$. In this case only the first two terms in equation (2.53) contribute and the DPNS intensity of spin-up (+) and spin-down (−) spin neutrons will be the same. Therefore if the intensities of spin-up and spin-down neutrons in the DPNS experiment are different, it is a signature of correlation between the structural and magnetic interfaces. Using DPNS techniques we have studied the separation and correlation of structure and magnetic interfaces for Ni film and it will be described in chapter 4 in detail.

Magnetic roughness is an important parameter which can influence the emerging magnetic properties at interfaces. The study related to defining magnetic roughness by considering two boundaries (magnetic and structure) at each interface described above requires an assumption of a homogeneous scattering potential along the in-plane directions (x, y) and has been successfully implemented for both PDNS [34] and x-ray resonant magnetic scattering [33, 35]. However, magnetic roughness can be defined in other ways such as in-plane magnetic domain distribution and inhomogeneities. Langridge *et al* [36] have explained the magnetic roughness by considering the orientation of magnetic domains, in the field dependence of diffuse neutron scattering on a Co/Cu multilayer system. More advancement in off-specular polarized neutron scattering with polarization analysis under DWBA [37–39] has been achieved in the last decade to study magnetic inhomogeneities in terms of magnetic domain structures in the plane of the film. Briefly, we will describe this approach in the next section.

2.7.2 Polarized DNS with spin polarization analysis under DWBA

The off-specular polarized neutron scattering from a magnetic in-plane domain distribution, ξ_x, becomes important if these are smaller than the coherence length, L_x^{coh}. If the domain length scales are either in the order of the coherence length or larger ($\xi_x \geqslant L_x^{\mathrm{coh}}$) an incoherent sum of intensities specularly reflected from individual domains are measured. However, if the domain size is smaller than the coherence area, neutrons are coherently reflected from the magnetic domains. The scattered intensity from all domains in the ith ellipsoid is a square of the sum of the amplitudes if the domain size is smaller than the coherence area and it can be written as $\left(\frac{d\sigma}{d\Omega}\right)_i^{\mathrm{coh}} = \left(\sum_j F_j\right)^2$, where F_j is the scattering amplitude for the jth domain. The scattering amplitudes from objects which are separated by a distance larger than the coherence volume add up incoherently, i.e. the total intensity is the sum of all scattering cross-sections indexed by the subscript, i, as $\frac{d\sigma}{d\Omega} = \sum_i \left(\frac{d\sigma}{d\Omega}\right)_i^{\mathrm{coh}}$.

Like specular reflectivity, off-specular scattering is also a coherent phenomenon of constructive interference from neutron waves scattered from different regions of the thin film within the coherence volume. Off-specular scattering from random in-plane interfaces has been described within the framework of the DWBA [9]. The details for polarized neutron off-specular scattering from in-plane domains has been given in [8, 37, 40, 41]. In general, within DWBA the lateral fluctuations are considered by a small correction or perturbation $V_{\mathrm{p}}(x, y)$ to the local and well-defined potential $V(z)$, which represents the depth dependence of the scattering potential averaged over the lateral coordinate and provides specular reflectivity. Thus the effective potential within DWBA is

$$V(r) = V(z) + V_{\mathrm{p}}(x, y). \tag{2.56}$$

Similarly, the neutron wave function ψ_{d} can be modified as

$$\psi_{\mathrm{d}}(r) = \psi_{\mathrm{l}}(r) + \psi_{\mathrm{p}}(r), \tag{2.57}$$

where $\psi_{\mathrm{l}}(r)$ is a scattered wave function from the laterally average region and can be treated with potential $V(z)$ for specular reflectivity. The scattered wave $\psi_{\mathrm{p}}(r)$ corresponds to the residual perturbation potential $V_{\mathrm{p}}(x, y)$. It is noted that both wave function and potential are neutron spin-dependent. Thus using the above-mentioned modified potential and wave function one can calculate the scattering amplitude in DWBA as described in [8, 40, 41]. The differential diffuse scattering cross-section is given by

$$\frac{d\sigma}{d\Omega} = \left(\frac{Ak_0^4}{16\pi^2}\right)\left|\sum_{n=1}^{N} C_n\right|^2 S(Q_x), \tag{2.58}$$

where C_n are matrix elements defined in [37] and lateral roughness structure factor $S(Q_x)$ for vertically correlated interfaces [41]:

$$S(Q_x) = \frac{\sigma_{\mathrm{m}}^2 \xi}{\sqrt{2\pi}} \exp\left(-0.5\frac{Q_X^2}{\xi^2}\right), \tag{2.59}$$

where σ_{m} and ξ are the magnetic roughness and average lateral correlation length (magnetic domain size in the lateral direction) at the interface.

2.8 Data analysis for reflectometry

It is a non-trivial task to obtain meaningful physical parameters for a thin film heterostructure from reflectometry data. Experimentally, one obtains the reflectivity data from a thin film heterostructure as a function of momentum transfer Q. In principle, one should be able to extract information in the real space regarding the system, through a Fourier transformation from momentum space (Q) to real space. But reflectometry experiments like most of the scattering techniques are done in limited Q space and direct Fourier transformation is not a viable option. Usually one resorts to model-fitting for extracting relevant information from experimental data. In this instance, one starts with a model of the given heterostructure in the form of a histogram as shown in figure 2.10(b), with the densities and magnetism in each slice as parameters of the model. Using the model, we simulate the neutron reflectivity and calculate the difference between experimental and simulated data using some error function E_{r} (e.g. χ^2 minimization). One may define a weighted error function χ^2:

$$\chi^2 = \sum_i \frac{[S_i^{\exp} - S_i^{\mathrm{Model}}]^2}{\mathrm{err}_i^2}, \tag{2.60}$$

where $S_i^{\exp}$ are the experimental data and S_i^{model} are the data generated using a model. The square of the difference between the two is weighted by the experimental error err_i^2. One modifies the model to reduce the error χ^2 between the experimental data and the model. The process of modifying the model comprises the search algorithm in parameter space. Model-fitting attempts to reach 'true' values of the parameters for the sample through a search in the parameter space. The model can accordingly be adjusted by some optimization method to obtain a closer agreement with the experimental data. The common problem of this method is that the deduced structure need not be the unique solution. A variety of data-fitting and parameter optimization strategies have been developed including *steepest gradient*, *downhill simplex*, or *conjugate gradient* etc [42]. The algorithms search for a minimum of χ^2 in the parameter space, which is the supposed global minimum, representing the true solution. A model histogram for the heterostructure is the starting point for such a search algorithm, where the set of physical parameter values in the histogram with slices is an assumed solution for the sample. If one makes N slices, then the parameter space is N-dimensional with a particular histogram as a solution.

All the search algorithms start with a single point in parameter space and the search proceeds point-by-point in this space. The authors of this book [4, 43] have adapted the technique of genetic algorithm (GA) for the optimization of reflectivity data (x-ray and neutron) from both specular and off-specular modes. The main advantage of this algorithm compared to traditional conjugate gradient methods is that the solutions are not easily trapped into a local minimum and one starts with a large population of possible solutions which makes them more efficient than other search algorithms. The approach takes advantage of the values of previous function evaluations. It is noted that none of the methods claims to derive a unique solution to the optimization problem. It is inherently impossible without *a priori* knowledge of the system. The choice of an appropriate error function is crucial for any data-fitting procedure regardless of the optimization method used. The GA gives us a great deal of flexibility in this choice, since we need only to choose a continuous function and do not require the function to have continuous derivatives. For fitting the neutron and x-ray reflectivity data the error function should have some additional properties such as (i) there should be a single deep global minimum and local minima, which are much less deep than the global minimum, (ii) it should be fast and simple to calculate, and (iii) it should have relative insensitivity to the absolute magnitude of the data since reflectivity data often span many orders of magnitude.

References

[1] Majkrzak C F 1996 Neutron scattering studies of magnetic thin films and multilayers *Physica* B **221** 342–56

[2] Felcher G P, Hilleke R O, Crawford R K, Haumann J, Kleb R and Ostrowski G 1987 Polarized neutron reflectometer: a new instrument to measure magnetic depth profiles *Rev. Sci. Instrum.* **58** 609–19

[3] Fitzsimmons M R and Majkrzak C 2005 *Modern Techniques for Characterizing Magnetic Materials* (New York: Springer) pp 107–55

[4] Singh S, Swain M and Basu S 2018 Kinetics of interface alloy phase formation at nanometer length scale in ultra-thin films: x-ray and polarized neutron reflectometry *Prog. Mater Sci.* **96** 1–50

[5] Ankner J F and Felcher G P 1999 Polarized-neutron reflectometry *J. Magn. Magn. Mater.* **200** 741–54

[6] Lekner J 1987 *Theory of Reflection of Electromagnetic and Particle Waves* (Leiden: Martinus Nijhoff)

[7] Zabel H 1990 X-ray and neutron scattering at thin films *Festkörperprobleme 30: Plenary Lectures of the Divisions Semiconductor Physics Thin Films Dynamics and Statistical Physics Magnetism Metal Physics Surface Physics Low Temperature Physics Molecular Physics of the German Physical Society (DPG), Regensburg, March 26–30* ed U Rössler (Berlin: Springer) pp 197–217

[8] Zabel H, Theis-Bröhl K and Toperverg B P 2007 Polarized neutron reflectivity and scattering from magnetic nanostructures and spintronic materials *Handbook of Magnetism and Advanced Magnetic Materials* (New York: Wiley)

[9] Sinha S K, Sirota E B, Garoff S and Stanley H B 1988 X-ray and neutron scattering from rough surfaces *Phys. Rev.* B **38** 2297–311

[10] Born M and Wolf E 1989 *Principles of Optics* (New York: Pergamon)

[11] Parratt L G 1954 Surface studies of solids by total reflection of x-rays *Phys. Rev.* **95** 359–69

[12] Névot L and Croce P 1980 Caractérisation des surfaces par réflexion rasante de rayons X. Application à l'étude du polissage de quelques verres silicates *Rev. Phys. Appl.* **15** 761–79

[13] Banu N, Singh S, Satpati B, Roy A, Basu S, Chakraborty P, Movva H C P, Lauter V and Dev B N 2017 Evidence of formation of superdense nonmagnetic cobalt *Sci. Rep.* **7** 41856

[14] Basha M A, Bhatt H, Kumar Y, Prajapat C L, Gupta M, Basu S and Singh S 2020 Annealing driven interface diffusivity in FePt/Cu multilayer *AIP Conf. Proc.* **2265** 030254

[15] Basha M A, Prajapat C L, Bhatt H, Kumar Y, Gupta M, Kinane C J, Cooper J F K, Langridge S, Basu S and Singh S 2020 Field dependent helical magnetic structure in a compensated Gd/Co multilayer *J. Magn. Magn. Mater.* **516** 167331

[16] Basha M A, Prajapat C L, Gupta M, Bhatt H, Kumar Y, Ghosh S K, Karki V, Basu S and Singh S 2018 Interface induced magnetic properties of Gd/Co heterostructures *Phys. Chem. Chem. Phys.* **20** 21580–9

[17] Bhatt H, Kumar Y, Prajapat C L, Kinane C J, Caruana A, Langridge S, Basu S and Singh S 2020 Emergent interfacial ferromagnetism and exchange bias effect in paramagnetic/ferromagnetic oxide heterostructures *Adv. Mater. Interfaces* **7** 2001172

[18] Singh S, Basha M A, Bhatt H, Kumar Y and Gupta M 2022 Interface morphology driven exchange interaction and magnetization reversal in a Gd/Co multilayer *Phys. Chem. Chem. Phys.* **24** 6580–9

[19] Singh S, Basu S, Bhatt P and Poswal A K 2009 Kinetics of alloy formation at the interfaces in a Ni–Ti multilayer: x-ray and neutron reflectometry study *Phys. Rev.* B **79** 195435

[20] Singh S, Basu S, Bhattacharya D and Poswal A K 2010 Physical and magnetic roughness at metal–semiconductor interface using x-ray and neutron reflectometry *J. Appl. Phys.* **107** 123903

[21] Singh S, Basu S, Gupta M, Majkrzak C F and Kienzle P A 2010 Growth kinetics of intermetallic alloy phase at the interfaces of a Ni/Al multilayer using polarized neutron and x-ray reflectometry *Phys. Rev.* B **81** 235413

[22] Singh S, Basu S, Vedpathak M, Kodama R H, Chitra R and Goud Y 2005 Ultrathin multilayer of Fe and Ge: structure and magnetic properties *Appl. Surf. Sci.* **240** 251–9

[23] Blundell S J and Bland J A C 1992 Polarized neutron reflection as a probe of magnetic films and multilayers *Phys. Rev.* B **46** 3391–400

[24] Marshall W and Lovesey S W 1971 *Theory of Thermal Neutron Scattering: The Use of Neutrons for the Investigation of Condensed Matter* (New York: Oxford University Press)

[25] Majkrzak C F 1989 Polarized neutron scattering methods and studies involving artificial superlattices *Physica* B **156–57** 619–26

[26] Basu S and Singh S 2006 A new polarized neutron reflectometer at Dhruva for specular and off-specular neutron reflectivity studies *J. Neutron Res.* **14** 109–20

[27] Radu F, Leiner V, Wolff M, Ignatovich V K and Zabel H 2005 Quantum states of neutrons in magnetic thin films *Phys. Rev.* B **71** 214423

[28] Als-Nielsen J and McMorrow D 2011 *Elements of Modern X-ray Physics* (New York: Wiley) pp 29–67

[29] Mandelbrot B B 1982 *The Fractal Geometry of Nature* (New York: Freeman)

[30] Schiff L I 1968 *Quantum Mechanics* (New York: McGraw-Hill)

[31] Holy V, Kubĕna J, Ohli´dal I, Lischka K and Plotz W 1993 X-ray reflection from rough layered systems *Phys. Rev.* B **47** 15896–903

[32] Singh S and Basu S 2006 Microscopic characterization of corrosion morphology: a study in specular and diffuse neutron reflectivity *Surf. Sci.* **600** 493–6

[33] Osgood III R M, Sinha S K, Freeland J W, Idzerda Y U and Bader S D 1999 X-ray scattering from magnetically and structurally rough surfaces *J. Magn. Magn. Mater.* **198–199** 698–702

[34] Singh S and Basu S 2009 Separation and correlation of structural and magnetic roughness in a Ni thin film by polarized off-specular neutron reflectometry *J. Phys. Condens. Matter* **21** 055010

[35] Lee D R, Sinha S K, Nelson C S, Lang J C, Venkataraman C T, Srajer G and Osgood R M 2003 X-ray resonant magnetic scattering from structurally and magnetically rough interfaces in multilayered systems. II. Diffuse scattering *Phys. Rev.* B **68** 224410

[36] Langridge S, Schmalian J, Marrows C H, Dekadjevi D T and Hickey B J 2000 Quantification of magnetic domain disorder and correlations in antiferromagnetically coupled multilayers by neutron reflectometry *Phys. Rev. Lett.* **85** 4964–7

[37] Rühm A, Toperverg B P and Dosch H 1999 Supermatrix approach to polarized neutron reflectivity from arbitrary spin structures *Phys. Rev.* B **60** 16073–7

[38] Toperverg B P 2001 Specular reflection and off-specular scattering of polarized neutrons *Physica* B **297** 160–8

[39] Radu F, Vorobiev A, Major J, Humblot H, Westerholt K and Zabel H 2003 Spin-resolved off-specular neutron scattering from magnetic domain walls using the polarized ^{3}He gas spin filter *Physica* B **335** 63–7

[40] Toperverg B P 2002 Off-specular polarized neutron scattering from magnetic fluctuations in thin films and multilayers *Appl. Phys.* A **74** s1560–2

[41] Nickel B, Rühm A, Donner W, Major J, Dosch H, Schreyer A, Zabel H and Humblot H 2001 Spin-resolved off-specular neutron scattering maps from magnetic multilayers using a polarized ^{3}He gas spin filter *Rev. Sci. Instrum.* **72** 163–72

[42] Press W H, Teukolsky S A, Vetterling W T and Flannery B P 1992 *Numerical Recipes in FORTRAN: The Art of Scientific Computing* 2nd edn (Cambridge: Cambridge University Press)

[43] Singh S and Basu S 2016 Simultaneous parameter optimization of x-ray and neutron reflectivity data using genetic algorithms *AIP Conf. Proc.* **1731** 080007

IOP Publishing

Neutron and X-ray Reflectometry
Emerging phenomena at heterostructure interfaces
Saibal Basu and Surendra Singh

Chapter 3

Understanding emerging phenomena at interfaces using specular neutron and x-ray reflectometry

3.1 Emerging phenomena at interfaces and characterization

Interfaces of materials are ubiquitous in Nature and they are present in all fields of science. Interfaces play important role in accomplishing the desired performance of materials that include physical, chemical, and mechanical properties. Interfaces dominate the behaviour of many types of materials ranging from traditional condensed matter to soft condensed matter, such as nanoparticles, thin solid films, colloids, and biological systems. The discovery of emergent magnetic phenomena at the interfaces of ultra-thin magnetic films is of both fundamental and technological interest and has been explored extensively in recent times. Emerging phenomena arise from the fact that translational and inversion symmetry is broken at the interfaces and discontinuity in atomic structure and consequently change in electronic band structure leads to an enormous number of effects at interfaces. Modification in the band structure and local coordination can create new magnetic states, including the creation of magnetic moments in nonmagnetic materials at the interfaces. The breaking of inversion symmetry at interfaces creates a unique axis perpendicular to the interface and induces a uniaxial anisotropy associated with the surface with an energy of $\sim -\boldsymbol{K}_s\,(\boldsymbol{S}_i \cdot \boldsymbol{n})^2$ [1], where $\boldsymbol{n}$, $\boldsymbol{S}_i$, and $\boldsymbol{K}_s$ are the interface normal, total spin, and anisotropy constants, respectively. For $\boldsymbol{K}_s > 0$, this energy produces perpendicular magnetic anisotropy (PMA), favouring a magnetization perpendicular to the interface. A large number of Co-based heterostructures with nonmagnetic heavy metals, Pt [2], Pd [3], Ir [4], Au [5] etc, were first studied for PMA, where the anisotropy of the interfacial orbital angular momentum was induced by the lowered symmetry. The interfacial orbital angular momentum of the transition metals was modified by the presence of heavy metals at the interfaces, which enhances its spin–orbit interaction, thereby increasing PMA [6]. The breaking of inversion symmetry at interfaces can also give rise to another important interfacial interaction, namely the Dzyaloshinskii–Moriya interaction (DMI) [7, 8], which plays an important role in understanding the non-collinear spin textures at interfaces as well as a possible reason for ferromagnetic behaviour being presented by antiferromagnetic heterostructures.

The interfacial DMI favours canting of spins toward each other around the direction perpendicular to the separation between them and the interface normal, thereby promoting non-collinear spin textures [1]. These are the two important interactions at interfaces that have contributed towards the understanding of many interfacial magnetic properties in magnetic thin films and heterostructures. In addition to these interfacial interactions, there exists another important phenomenon, known as exchange bias [9, 10], at the interface between ferromagnets (FMs) and antiferromagnets (AFMs) caused by antiferromagnetic coupling of spins at an interface. Exchange bias shifts the hysteresis loop of the ferromagnet in one field direction, resulting in unidirectional anisotropy. Recently exchange bias phenomena in many oxide heterostructures between AFM–AFM, FM–FM, and other magnetic systems have emerged due to different interactions and will be discussed in detail later in this chapter.

The artificial arrangements of various magnetic elements or compounds in a thin film generate a rich environment for studying the effect of interfaces, symmetries, and interactions on their magnetic properties. Multilayers as a combination of multiple interfaces in magnetic thin films are other important nanomagnetic systems that provide a variety of important effects, e.g. giant magnetoresistance (GMR) [11], interlayer exchange coupling (IEC) [12], non-collinear coupling, etc. As discussed in chapter 2, periodic multilayers, which increase the probing volume and give rise to the so-called Bragg reflection due to their artificial periodicity, are often preferable to single or bilayer thin films. Bragg reflections represent the Fourier transform of the real periodic structure and the reflectivity from multilayers is much more sensitive to small perturbations than reflectivity from single films. XRR and PNR are complementary techniques to characterize multilayers. In addition, PNR provides layer-resolved vector magnetometry in the entire thickness of the multilayer (from top to bottom). The surfaces and interfaces in magnetic multilayer systems induce many phenomena attributed to symmetry breaking, loss of coordination number, finite thickness, atomic and electronic reconstructions, exchange interaction, and proximity effects. These effects are intrinsic in the multilayer systems and they would occur even at ideal and defect-free interfaces. However, ideal interfaces between defect-free materials do not generally exist. Sometimes lattice mismatch and the consequent strain at interfaces in a heterostructure produce defects that are intrinsic to the heterostructure. The intrinsic defects and disorders in combination with extrinsic effects in magnetic films and heterostructures, such as interdiffusion, roughness, and alloying at interfaces, can induce and even control novel emergent phenomena in magnetic heterostructures. Thus it requires sensitive characterization tools to enable advanced investigation of emerging phenomena at interfaces (importantly buried interfaces) in relation to their structure with inherent good resolution. PNR is an ideal non-destructive tool to reveal the emergent magnetic properties of buried interfaces.

Thin films and multilayer (nanostructure) systems are being investigated intensively due to their novel properties and promise for possible technological applications. The key issue for these materials is to understand the correlation between their physical properties (transport, magnetism, mechanical, etc) and their chemical–physical structures, particularly at interfaces. In principle, detailed knowledge of the chemical and physical structures allows the calculation of their physical properties. Magnetic thin films have been of great interest due to their technological application in magnetic devices. The physics and properties of these thin films and multilayers cannot be understood without a deep insight into the structure of their surfaces and interfaces. PNR is a non-destructive and well-developed technique that can determine the chemical and magnetic structure at the nanometre length scale in thin films and heterostructures. The neutron is neutral, it has a large scattering length for low atomic number materials unlike x-rays, and it possesses a magnetic moment of $-1.91\ \mu_N$. These properties make PNR an attractive tool, complementary to x-rays, for studying thin films of polymers and magnetic materials.

Apart from studies related to magnetic properties at the interfaces, it is also important to understand alloying at the interfaces of multilayers as this phenomenon provides intriguing physics at interfaces. Interface alloy formation in ultra-thin films (tens of nanometres) and its kinetics is important for interface magnetic properties in these systems. The authors have carried out detailed work on the characterization of interface alloys and their magnetic properties in Ni/Ti [13] and Ni/Al multilayers [14–16]. The authors also used PNR and XRR as two independent measurements to quantify binary alloy compositions at interfaces [13, 17] from the respective fitted densities. We have provided a discussion on interface alloying and magnetic properties of interface alloys, using PNR and XRR, later in this chapter.

As mentioned in chapter 2, the general principle of a PNR experiment is illustrated in figure 3.1, in which a polarized neutron beam with polarization P collinear to the applied magnetic field (H) falls onto the magnetic thin film surface at an angle of incidence θ_i. The in-plane magnetization (M) of the film is at an angle α with the applied field. The spin-dependent PNR [14, 17–19] is defined by four reflectivities depending on the polarization channel, R^{++}, R^{--}, R^{+-}, and R^{-+}, where the first and second superscript denotes the direction of the incoming and reflected neutron polarization as parallel (+) or antiparallel (−) with respect to the external guide field direction. PNR with no change in the polarization upon reflection is termed as non-spin-flip (NSF) reflectivities (i.e. R^{++}, R^{--}). The two non-spin-flip reflectivities, R^{++} and R^{--}, are related to the magnetization components parallel ($M_{||}$) to the applied (in-plane magnetization of the sample) field. Whereas PNR with the change in the polarization of the reflected beam is termed as spin-flip (SF) reflectivities (i.e. R^{+-} and R^{-+}) and are related to the magnetization components perpendicular ($M_{\perp}$) to the applied field. Thus employing all spin-dependent neutron reflectivity with polarization analysis provides the direction of in-plane magnetization (magnitude of M and angle of rotation α of magnetization with respect to the applied field) along the depth of the heterostructures. For a depth-dependent magnetic structure, one measures specular reflectivities (angle of incidence θ_i = angle of reflection $\theta_f = \theta$). Therefore, the resultant momentum transfer (Q) is

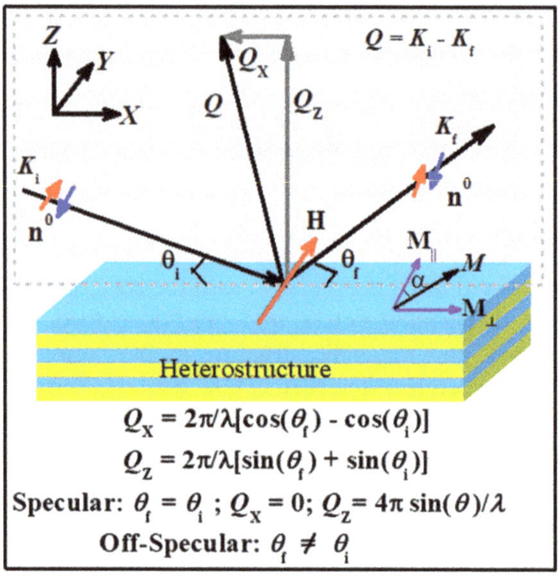

Figure 3.1. Schematic of PNR experiment with a polarized neutron beam. Neutron beam incident (wave vector $\boldsymbol{K}_i$) at the surface of the film with an angle of incidence of θ_i and reflected (wave vector $\boldsymbol{K}_f$) at an angle of reflection of θ_f. The difference between the incoming and outgoing wave vectors is defined as momentum transfer vector $\boldsymbol{Q}$ (i.e. $\boldsymbol{Q} = \boldsymbol{K}_i - \boldsymbol{K}_f$). The polarization of the neutron, P, is shown in the plane of the sample parallel to the applied magnetic field (H). The magnetization M (at an angle α with H) of the thin film is shown with the two in-plane components $M_{||}$, parallel to H, and $M\perp$, perpendicular to H.

equivalent to the momentum transfer component Q_z normal to the sample surface and is given as $Q = Q_z = \frac{2\pi}{\lambda}[\sin(\theta_f) + \sin(\theta_i)] = \frac{4\pi}{\lambda}\sin(\theta)$, where λ is the wavelength of the neutron. Off-specular reflectivity ($\theta_i \neq \theta_f$) originates from lateral structures, such as interfacial roughness and magnetic domains, which break the in-plane translational symmetry of the sample and lead to an in-plane momentum transfer $Q_x = \frac{2\pi}{\lambda}[\cos(\theta_f) - \cos(\theta_i)]$.

Experimentally, specular PNR data can be measured in two ways. (i) When the angle ($\alpha = 0$) of rotation of magnetization is zero, this case arises when the applied magnetic field is sufficient to align the magnetization of the film along the applied field and this measurement can be termed as PNR without polarization analysis. The measured spin-dependent neutron reflectivity, in this case, is denoted by spin-up (R^+) and spin-down (R^-) reflectivities and specular PNR measurement provides the depth-dependent magnetization information of the thin-film systems. (ii) when α is finite, in this case, all four reflectivities (R^{++}, R^{--}, R^{+-}, and R^{-+}) contribute and the measurement is termed as PNR with polarization analysis. Measuring all four reflectivities in the case of specular PNR with polarization analysis provides detailed depth profiles of magnetic structure (magnetic moment and its direction with respect to applied magnetic field). In this chapter, we will discuss different interfaces and associated phenomena investigated primarily using specular PNR. Specular XRR also provides the depth-dependent layer structure of the thin-film system.

3.2 Ferromagnetic metal/semiconductor heterostructures

The discovery of the GMR effect observed in a multilayer of alternating layers of magnetic and nonmagnetic materials in the early 1980s [11] led to the development of first-generation spintronics materials, combining electrons' spin degrees of freedom with charge. The possibility of controlling the electron's spin in semiconductors was first demonstrated by Kikkawa and Awschalom [20] and a large spin coherence length of ~100 μm was found in the semiconductor. The long spin coherence length in semiconductors establishes the possibility of constructing the next generation of spintronics by integrating spin with a conventional semiconductor device. Two classes of material systems exploited for use in these spintronic devices are (i) dilute magnetic semiconductors (DMS) or magnetic metal-doped III–V and II–VI semiconductors such as GaMnAs, CdCoSe, etc, and (ii) ferromagnetic/semiconductor (FM/S) heterostructures, such as Fe/GaAs, Co/Si, Fe_3O_4/GaAs, CoMnGa/GaAs, Fe/Ge, etc. The DMS have the advantage of integrating with semiconductor materials and devices, however, the challenge is to increase their Curie temperature (T_c) [21] for application. The highest T_c obtained so far in DMS is around 150 K. The lattice constant of bcc Fe is 2.866 Å, which is about half of the lattice constant (~5.654 Å) of semiconductor GaAs and this has made it possible to grow the first epitaxial heterostructure of Fe/GaAs [22]. FM metal–semiconductor interfaces display exotic physical and chemical properties [23, 24] and have become a focus of intensive studies for their structure and novel magnetic properties. These systems have also stimulated many interface-driven experimental and theoretical studies [25–28]. Following the work of Datta and Das [29], a new generation of electronic devices (spintronics), exploiting the spin degree of freedom of carriers in FM/S interfaces [30], have been envisaged, namely spin transistors, spin memories, spin field-effect transistors, etc [31]. The conductivity mismatch between the FM and semiconductor poses a huge problem for the injection of spin-polarized electrons from the FM to semiconductor [32]. FM/S heterostructures, e.g. 3*d* transition metal FMs (Fe, Co and Ni, etc) on semiconductors (Si, Ge and GaAs, etc), are promising systems for the transfer of magnetic or spin-related functions into semiconductor materials, a necessity for the realization of spintronic devices [30, 33]. However, nonmagnetic alloy formation at the interfaces of the FM/S heterostructures has been detrimental to the electron spin injection from the FM to the semiconductor. Therefore, the property of interest, i.e. the transport of spins across the FM/S interface, will strongly depend on interface properties, namely roughness, alloying at interfaces, and interfacial magnetic moment. PNR is an excellent tool to simultaneously probe the interface magnetization and structure in such important heterostructures. Here, we will discuss the interface magnetization of Fe–Ge multilayer systems in detail.

The Fe/Ge-based FM/S heterostructure is one of the systems investigated extensively for possible application in the field of spintronics. Ge has a bandgap of 0.65 eV at 300 K and shows only a 1.3% lattice mismatch with *bcc* Fe. Ge also shows high electron–hole mobility with excellent compatibility with Si-based technology [34]. Fe/Ge heterostructures also exhibit a desirable room temperature (RT) interfacial ferromagnetic property. Therefore the RT ferromagnetic intermetallic alloys (germanides) formed at the interface of Fe/Ge heterostructures may facilitate electron spin injection from the FM to the semiconductor [35] for controlled manipulation of the charge and the spin of electrons in FM/S systems. In addition, germanide alloys can serve as interconnects for Ge-based nanoelectronics [36], which may improve device performance by reducing interface defects and contaminations, as well as by eliminating possible interface oxides. However, several theoretical analyses using epitaxial Fe/Ge supercell structures indicated that the Fe moment reduces at the interfaces [37, 38]. The spin-resolved electronic band structure calculations [37, 38] suggested that reduced atomic coordination, modified volume per atom, and interface-induced electronic states strongly affect the magnetic moment of iron at the Fe/Ge interface. Interfacial magnetization in this system also depends strongly on the crystalline state of Fe and Ge as well as the deposition sequence. X-ray photoelectron spectroscopy (XPS) showed that when crystalline Fe is deposited on amorphous Ge, the semiconductor penetrates deeper inside the iron layer compared to the situation when the amorphous semiconductor is deposited on Fe [39]. Singh *et al* [40–43] investigated several Fe/Ge heterostructures using XRR and PNR techniques and correlated the structure and magnetic properties at interfaces.

3.2.1 Asymmetric and reduced interfacial magnetization in the Fe/Ge multilayer

To investigate interface magnetization of the Fe/Ge system at room temperature (RT), Singh *et al* grew an Fe/Ge multilayer with ten repetitions of the Fe/Ge bilayer on a Si (substrate)/Ge (buffer layer) heterostructure [41]. The inset of figure 3.2 shows the schematic of the multilayer, which was grown by RF sputtering. The nominal thickness of the Fe and Ge layers were 75 and 55 Å, respectively. The multilayer was characterized using a combination of techniques including XRR, XRD, SQUID

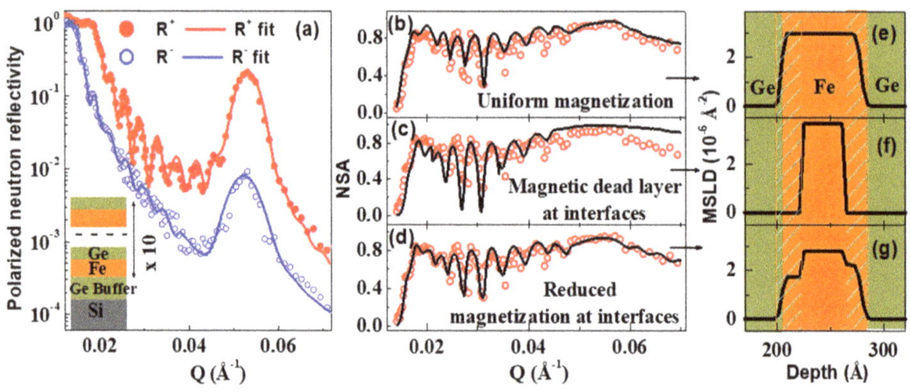

Figure 3.2. (a) Spin-dependent polarized neutron reflectivity (R^+ and R^-) data (symbols) and corresponding best fits (solid lines) for an Fe/Ge multilayer. The inset shows the schematic of the Fe/Ge multilayer (repetition of ten bilayers) grown on Si (substrate)/Ge (buffer layer). Panels (b)–(d) show the normalized spin asymmetry (NSA) = $(R^+ - R^-)/(R^+ + R^-)$ data (open circles) and the fits (solid lines) corresponding to three different magnetization models shown in (e)–(g). The shaded regions in magnetization depth profiles in (e)–(g) show the interfacial region at two interfaces. NSA fits for corresponding models are shown by the arrow in the figure. Adapted with permission from [41]. Copyright 2007 AIP Publishing.

magnetometry, and PNR. The XRD results suggested the growth of polycrystalline *bcc* Fe and amorphous Ge layers. Specular XRR measurements provided a detailed depth profile of electron scattering length, which validate the multilayer structure. The macroscopic magnetic properties of the multilayer at RT and low temperature were studied using SQUID magnetometry and the results at RT suggested a reduced average magnetic moment of 1.17 μ_B/Fe atom. XRR and unpolarized neutron reflectometry (NR) were used to obtain the electron scattering length density (ESLD) and nuclear scattering length density (NSLD), respectively. PNR measurements at RT were performed to explore the magnetization depth profiles of the multilayer with special emphasis on estimation of the interfacial magnetization. The Fe/Ge multilayer structure, obtained from XRR and unpolarized neutron reflectivity, was used as the input to fit spin-dependent PNR data (R^+ spin-up and R^- spin-down) shown in figure 3.2(a). However, a uniform magnetization for the whole Fe layer could not fit the PNR data and thus different magnetization models were considered to fit the PNR data. The spin-dependent PNR data can be represented as normalized spin asymmetry (NSA = $(R^+ - R^-)/(R^+ + R^-)$) for highlighting the fits for different magnetization models. The magnetic scattering length density (MSLD) or magnetization models considered to fit the PNR (or NSA) data are shown in figures 3.2(e)–(g), which show the magnetization depth profile across two interfaces, i.e. Fe/Ge (Ge layer grown on Fe layer) and Ge/Fe (Fe layer grown on Ge layer), of an Fe/Ge bilayer. The three models are (i) a uniform magnetization for each Fe layer, (ii) zero magnetization for the interfacial Fe layer with uniform magnetization for the middle part of each Fe layer, and (iii) reduced interfacial magnetization with uniform magnetization for the middle part of the Fe layer. These models are depicted in figures 3.2(e), (f), and (g), respectively. The corresponding fits to PNR data for these models are shown in figures 3.2(b), (c), and (d). It is evident from figure 3.2 that reduced interfacial magnetization with asymmetric distribution for two interfaces (figure 3.2(g)) fitted the PNR data (figures 3.2(a) and (d)) best. The asymmetric nature of magnetization at two interfaces was attributed to the asymmetric interdiffusion of atoms at interfaces due to different nucleation and growth of crystalline Fe and amorphous Ge layers at two interfaces [41]. The study also suggested a reduction in the magnetic moment of Fe (~1.30 μ_B) in the middle bulk region of the Fe layer compared to the bulk, which further reduces to 1.04 and 0.81 μ_B near two interfaces [41]. The thickness-average magnetization at RT for Fe from PNR data closely matches the average magnetic moment obtained from SQUID results. The results also suggested that controlling interdiffusion or interface roughness during the growth of the multilayer will enhance the magnetic properties at the interfaces. However, the reduction in magnetization of the Fe layer near interfaces will be detrimental to the spin injection across the interfaces.

3.2.2 Different structural and magnetic roughnesses at the interfaces of Fe/Ge heterostructures

Using the intrinsic properties of neutrons, one can simultaneously obtain both the nuclear and magnetization depth profiles of the heterostructures [17, 18, 44].

Combining specular XRR, off-specular XRR, and PNR measurements at RT, Singh *et al* [42] studied a Ge/Fe/Ge trilayer heterostructure and investigated the interface-dependent magnetization in this system. The inset of figure 3.3(a) shows the schematic of the structure of the Ge/Fe/Ge heterostructure grown on the Si single-crystal substrate. XRD data from the heterostructure suggested the formation of crystalline *bcc* Fe and amorphous Ge layers. The spin-dependent PNR data (symbols) at RT from the heterostructure, together with the fits (solid lines) are shown in figure 3.3(a). The corresponding NSA $((R^+ - R^-)/(R^+ + R^-))$ data (symbols) and fit (solid line) from the heterostructure are shown in figure 3.3(b). Specular XRR and PNR indicated commensurate depth profiles of electron and nuclear scattering length density (ESLD and NSLD), respectively. Figure 3.3(c) shows the NSLD (black solid line) and MSLD (red solid line) depth profile of the

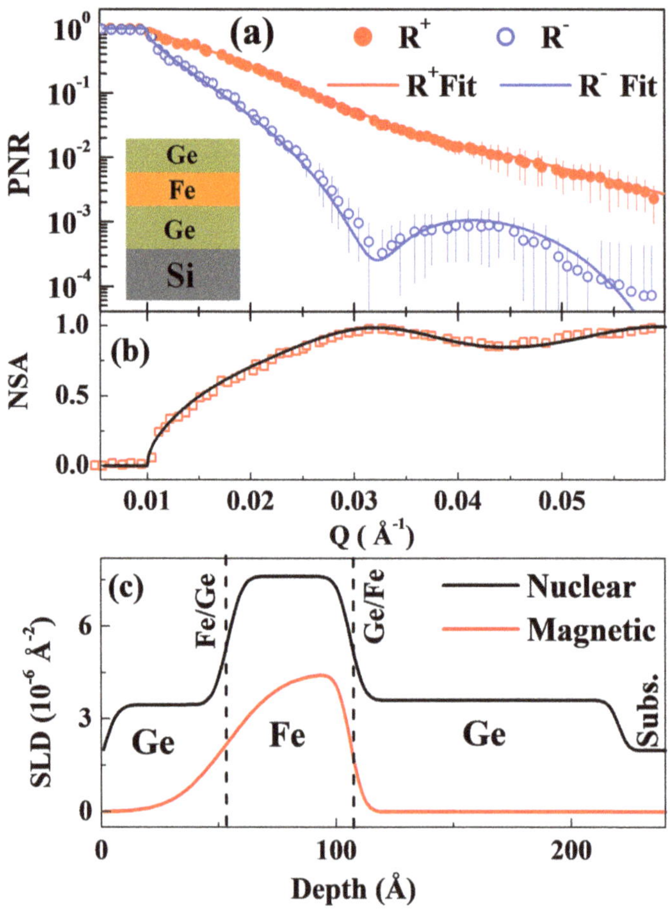

Figure 3.3. (a) Polarized neutron reflectivity (PNR) data (symbols) and corresponding best fit (solid lines) for the Ge/Fe/Ge heterostructure grown on a Si substrate. The inset shows the schematic of the heterostructures. (b) Normalized spin asymmetry (NSA) data (symbols) and best fit (solid line). (c) Nuclear and magnetic SLD depth profiles for the heterostructures which best fitted the PNR data. The vertical dashed lines indicate two interfaces across the Fe layer. Adapted with permission from [42]. Copyright 2010 AIP Publishing.

heterostructure obtained from PNR. Specular reflectivities (both XRR and PNR) provided thicknesses of 115, 58, and 46 Å, for the Ge, Fe, and Ge layers, respectively, thus confirming the layer structure of the heterostructure as Si (substrate)/Ge (115 Å)/Fe (58 Å)/Ge (46 Å).

Specular reflectivity also suggested a structural roughness of 10 Å for both the Ge/Fe (Fe grown on Ge) and Fe/Ge (Ge grown on Fe) interfaces. However, off-specular XRR measurements suggested different amounts of interdiffusion at these interfaces due to different kinds of nucleation at two interfaces (i.e. crystalline layer on the amorphous layer and vice versa) [42]. The PNR data suggested a drastically different magnetization depth profile (the red line in figure 3.3(c)) at two interfaces, suggesting that the magnetic roughness is higher at the interface where the interdiffusion of atoms is larger. The study also suggested asymmetric magnetization profiles due to the nucleation and growth of layers with different atomic structures (i.e. amorphous and crystalline nature). A rough magnetic interface was observed when an amorphous Ge layer was grown on a crystalline Fe layer despite showing the same structural roughness at the two interfaces.

Thus the study demonstrates the capability of x-ray and PNR techniques in revealing the details of physical and magnetic interfaces on a microscopic length scale. The findings have strong bearing on the characterization of interfaces in metal–semiconductor-based thin films for device applications. The combination of two non-destructive techniques (XRR and PNR) in both specular and off-specular modes reveals the depth-dependent structure and magnetization properties as well as interface morphology of different interfaces even for buried interfaces.

3.2.3 Formation of nanocrystalline alloy phase with large magnetic anisotropy at interfaces on annealing of the Fe/Ge multilayer

In another study on the Fe/Ge multilayer, Singh *et al* [43] studied the effect of interface morphology, which was varied by controlled annealing of the Fe/Ge multilayer, on the interfacial magnetic properties of the system. It was important to investigate the emerging FM interfacial alloy layer which may be of help to transfer spin across the FM/S interface in this system for possible application in spintronics [30]. The Fe/Ge multilayer studied was a repeat of five bilayers of Fe (thickness ~50 Å) and Ge (thickness ~30 Å) grown on Si (substrate)/Ge (buffer layer of thickness ~150 Å) heterostructure and an Au capping layer (25 Å) at the air–film interface. The schematic of the multilayer is shown in the inset of figure 3.4(a). The Fe/Ge multilayer was annealed at different temperatures in a vacuum and characterized *ex situ* using different techniques, e.g. XRD, XRR, SQUID, and PNR. Figure 3.4(a) shows the XRD data from the multilayer on annealing at different temperatures. The XRD data suggest the formation of crystalline Fe layers and amorphous Ge layers, which remain intact on annealing the multilayer below 573 K. The multilayer annealed at 573 K shows a Bragg peak for an alloy (Fe_3Ge) phase and no peaks corresponding to Fe and Ge were observed. XRR measurements (figure 3.4(b)) show Bragg peaks up to many orders as a result of the periodicity of Fe and Ge layers in the pristine sample. Modification in XRR data, such as reduction in the reflectivity

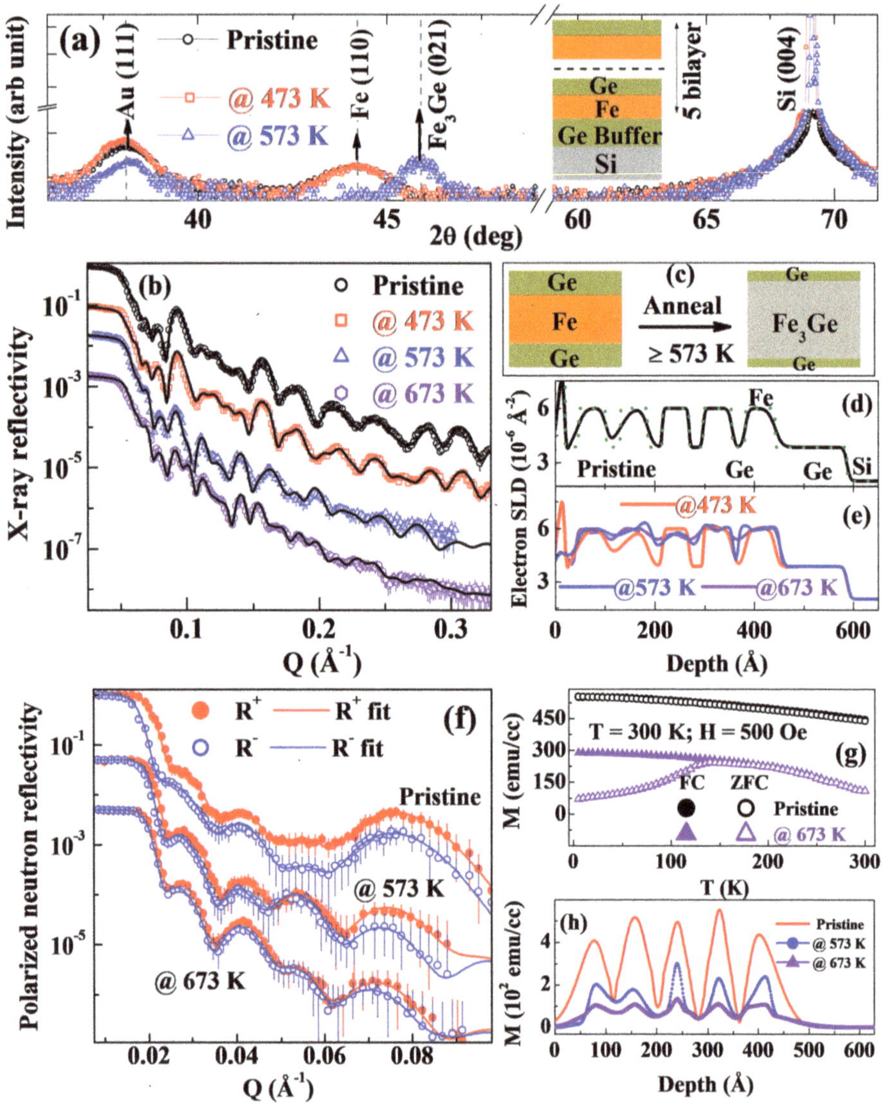

Figure 3.4. (a) XRD data for the Fe/Ge multilayer in pristine and annealed (different temperatures) conditions. Different Bragg peaks are indicated. The inset shows the schematic of the structure of the Fe/Ge multilayer. (b) XRR measurements for pristine and multilayers annealed at different temperatures, which are shifted vertically for better visualization. (c) Schematic of the Fe/Ge bilayer across two interfaces before and after annealing the multilayer at a temperature $T \geqslant 573$ K. Electron scattering length density (ESLD) depth profile for (d) pristine and (e) annealed Fe/Ge multilayers, which best fitted the corresponding XRR data. The ELSD profiles for annealed samples suggested the formation of an alloy layer at the interfaces. (f) PNR data from the multilayer for different conditions of annealing and shifted vertically for better visualization. (g) SQUID data from the multilayer. (h) Magnetization depth profiles as obtained from PNR data of the multilayer at different conditions. Reproduced with permission from [43]. Copyright 2021 Elsevier.

and shift in Bragg peak position, etc, for an annealed multilayer at $T \geqslant 573$ K clearly suggest alloy formation at the interfaces, which is shown in figure 3.4(c). A layer structure model (figure 3.4(c)) for the heterostructure has been used for fitting XRR and PNR data for different temperatures of annealing. The electron scattering length density (ESLD) depth profiles of the multilayer under different annealing conditions are shown in figures 3.4(d) and (e). After annealing at 673 K, the Fe layers were consumed to form an alloy/compound Fe_3Ge. This is evident from the electron density profile in figure 3.3(e).

A similar modification in PNR data for annealed multilayers can be seen in figure 3.4(f) where we have plotted (vertically shifted for different annealing conditions) spin-dependent PNR (R^+ spin-up and R^- spin-down) data for pristine and multilayers annealed at different temperatures. Formation of the alloy layer is also seen in the x-ray diffraction pattern, as a (021) Fe_3Ge peak (figure 3.4(a)). The magnetization depth profiles for the multilayers at different annealing conditions are shown in figure 3.4(h). The NSLD depth profile (not shown here) at different annealing conditions shows a similar trend as shown by the corresponding ESLD depth profiles obtained from XRR. The XRR and PNR results showed the formation of alloy layers at the interfaces. It is evident from the magnetization profiles of the annealed multilayers that the alloy phase shows long-range magnetic order (FM) with a transition temperature above RT. The micromagnetic measurements (the SQUID data in figure 3.4(g)) also suggest modification in the field cooled (FC) and zero-field cooled (ZFC) data on annealing the multilayer. The splitting of FC and ZFC data for the annealed multilayer indicates magnetic disorder due to the formation of magnetic nanoparticles of the alloy.

This study suggested that the annealing of the multilayer at a temperature of $T \geqslant 573$ K resulted in the formation of a nanocrystalline alloy phase as a consequence of the large diffusivity of elements at the interfaces, which is 4 (~2) orders larger in magnitude than the self-diffusivity of crystalline Ge (Fe). The diffusivity of the atoms at interfaces was measured by tracking the change in reflectivity at the Bragg peak [17]. Interestingly the alloy layer formed at a much lower temperature (~523 K) than the temperature of formation of the alloy in bulk (~973 K). Formation of the alloy phase on annealing the multilayers is also indicated by the evolution of nanoparticles with a high magnetic anisotropic constant. The study further suggested that the emerging alloy phase formed at relatively low temperatures by annealing the Fe/Ge multilayer is structurally stable at higher temperatures and compatible for use in semiconductor technology. PNR measurements suggested RT ferromagnetism for the alloy phase with reduced average magnetization. Thus the results indicated that the RT ferromagnetic alloy may be the perfect candidate for high-quality magnetic contacts, interconnectors, and interfaces in the field of spintronics. This may facilitate electron spin injection from the FM to the semiconductor for controlled manipulation of the charge and the spin of electrons in FM/S systems. This work presented on Fe/Ge systems highlights the importance of PNR (and XRR) in quantifying the formation of alloy layers at the interfaces, together with their physical and magnetic characteristics with an excellent spatial resolution.

Ni/Ge is another important FM/S system, which has motivated many experimental and theoretical studies in the field of magnetism and nanotechnology. Ni–germanides are an important class of metal–semiconductor alloys because of their suitability in microelectronics applications. Ni–germanides are also highly attractive as a promising germanide material for Ge-based semiconductor technology because of their low processing temperature (~270 °C) as compared to other transition metal germanides. Low processing temperature is effective to prevent thermal degradation of the gate material of the germanides. The Ni–Ge binary system shows many phases and thermal annealing is one of the suitable methods for inducing solid-state reactions between Ni and Ge with short bilayer periods to produce nickel germanides [45]. Low resistivity is observed for the NiGe alloy phase as compared to that of Ni-rich germanides. Using XRR and PNR, Swain *et al* [45] studied Ni/Ge multilayers and observed the formation of Ni–germanide alloy at interfaces. Similar to the Fe/Ge systems described earlier, the pristine Ni/Ge multilayer also showed polycrystalline growth of Ni layers and an amorphous phase for the Ge layers. To study the effect of interface morphology on the structure, and magnetic and transport properties of this system, the multilayer was annealed at 250 °C in vacuum for different time intervals of 0.5, 1.5, and 4 h. The transport (resistance) measurement showed a metallic trend over the entire temperature range of (10–300 K) and the value of electrical resistance for the annealed sample was lower than the resistance of the pristine multilayer over the entire temperature range. XRR and PNR results suggested the formation of an alloy phase at the interfaces in this system and the composition of the alloy phase was estimated using reflectivity (XRR and PNR) techniques. The alloy phase was found to be a NiGe alloy. However, the PNR results suggested that the alloy phase is not ferromagnetic. A combination of the transport and reflectivity results suggested that the alloy NiGe phase is contributing to the observation of lower electrical resistance on annealing the Ni/Ge multilayer [45].

3.3 Interlayer exchange coupling: ferromagnetic metal/nonmagnetic metal heterostructures

3.3.1 Bilinear exchange coupling

Starting in 1986 various groups around the world discovered the interlayer exchange coupling (IEC) between two FM layers separated by a nonmagnetic (NM) interlayer metal layer [12, 46]. It was found that the IEC across most 3*d*, 4*d*, and 5*d* NM metallic spacer layers oscillate between AFM and FM as a function of spacer layer thickness. This discovery enabled the control of AFM coupling between two FM films and it is now used in most spintronic devices [47]. This collinear (AFM (180°) and FM (0°)) coupling also depends on the interlayer material and its thickness (figure 3.5). Figure 3.5 illustrates the schematic of the oscillatory exchange coupling in an FM/NM heterostructure, where J is the coupling parameter. Phenomenologically, the bilinear exchange coupling energy per unit area (E) between two FM layers 1 and 2, separated by an NM interlayer of thickness, t (figure 3.5), can be given by

$$E = -J_{\mathrm{L}}\cos(\Delta\theta), \tag{3.1}$$

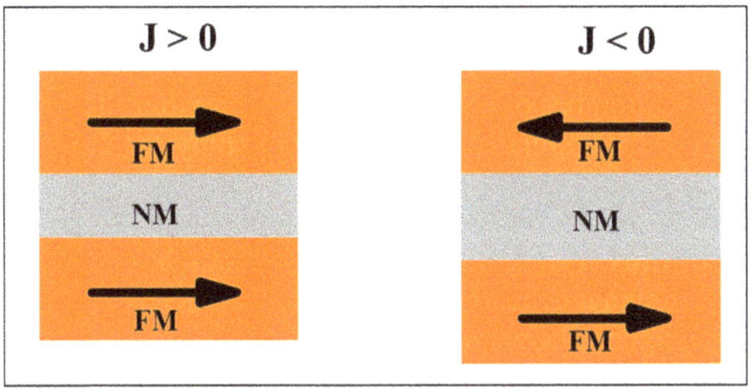

Figure 3.5. Schematic illustration of the oscillatory exchange coupling in ferromagnetic (FM)/nonmagnetic (NM) heterostructure. J is the coupling parameter.

where J_L is the bilinear coupling constant and $\Delta\theta$ is the angle between the magnetization of two FM layers. The value of J_L decides the nature of coupling with $J_L > 0$ (positive) and $J_L < 0$ (negative) suggesting FM and AFM coupling, respectively. The coupling is called bilinear because the energy per area is linear in the directions of both magnetizations.

Magnetic multilayers and superlattices have been extensively studied after the discovery of AFM coupling and IEC in Fe/Cr superlattices [11, 46] and the associated GMR effect [48, 49], which is induced by spin-dependent scattering of the conduction electrons. Later a large number of 3*d*, 4*d*, and 5*d* transition metal (TM) systems were found, which showed the oscillatory exchange coupling [50]. To understand this collinear oscillatory coupling two general approaches have been employed. (i) The phenomenological indirect exchange interaction called the Ruderman–Kittel–Kasuya–Yoshida interaction (RKKY theory), was originally developed to explain the oscillatory spin-polarization behaviour of dilute magnetic clusters in a nonmagnetic host. (ii) The other method is based on consideration of the quantum confinement effect arising from the periodic modulations of the superlattice structures and thus the formation of quantum well states (QWSs) within the nonmagnetic spacer.

The RKKY interaction is indirect interaction that is mediated by conduction electrons and it is oscillatory between AFM and FM with an oscillation period of $\lambda_f = \pi/\boldsymbol{k}_f$, where $\boldsymbol{k}_f$ is the wave vector of the spherical Fermi surface for the case of free electrons in the interlayer in the limit of large interlayer distances [51]. However, λ_f was found to be smaller than the oscillation periods for FM/NM/FM systems. In 1991 this behaviour was first explained by Bruno and Chappert [52] and Coehoorn [53] by considering the topological properties of the Fermi surface of the interlayer and the discrete atomic structure of the material within an RKKY approach. These studies described the period of oscillation, the temperature dependence, and the magnitude of the interlayer coupling very well for a large number of systems. Thus the Bruno and Chappert approximations [52, 54, 55] dealt with the real Fermi surfaces of Cu, Ag, and Au spacers and predicted oscillatory behaviour, whereas in the case of the original RKKY approach they correlated the oscillatory exchange coupling with nesting vectors which connect parts of the Fermi surface of the interlayer. The accurate quantitative predictions of Bruno and Chappert immediately led to a number of experimental activities and the Co/Cu heterostructure turned out to be a popular model system to study the IEC effect because of achieving epitaxial growth of this heterostructure easily.

Similar results of oscillating exchange interaction were obtained, at about the same time, by Edwards *et al* [56] and their model was based on the formation of QWSs within the nonmagnetic spacer instead of an RKKY approach. The FM alignment of two magnetic layers, where all majority spin bands lie below the Fermi energy (E_F) and thus electrons of this spin are confined to the nonmagnetic interlayer, form the QWSs. Such confinement of spin causes a discretization of states and it becomes more closely spaced with increasing spacer thickness and it gives rise to an oscillatory dependence on the band structure energy as the new states move below E_F thus causing the oscillatory exchange coupling.

3.3.2 Biquadratic exchange coupling

In 1991 Rührig *et al* [57], using Kerr microscopy and hysteresis curves measured using the magneto-optic Kerr effect (MOKE), reported a non-collinear (magnetization aligned at 90°) coupled magnetization profile in addition to collinear 180° (AFM) and 0° (FM) coupling for Fe/Cr/Fe heterostructures. They observed FM and AFM coupling between the top and bottom FM layers at small and large Cr interlayer thicknesses, respectively, whereas in a transition region they found that the top and bottom layers are oriented at 90° relative to each other. This indicates that in this Cr thickness regime an exchange coupling exists that favours a 90° orientation over an FM or AFM alignment even in an applied field that wants to align all Fe layers parallel. In a phenomenological approach, Rührig *et al* [57] defined their finding by suggesting the existence of an additional biquadratic exchange coupling term J_Q which can favour a spin alignment as opposed to the usual bilinear Heisenberg term J_L which can cause the AFM alignment. The biquadratic exchange coupling energy per unit area can then be written as

$$E = -J_Q \cos^2(\Delta\theta). \tag{3.2}$$

The effective exchange coupling energy per unit area is

$$E = -J_L \cos(\Delta\theta) - J_Q \cos^2(\Delta\theta). \tag{3.3}$$

It can be seen easily that a competition between the first and second terms in equation (3.3) can lead to an arbitrary angle ($\Delta\theta$) between magnetic moments in the range of 0°–180° and not only parallel, antiparallel, or 90° ordering of magnetic moments. In the absence of an external magnetic field and magnetic anisotropy, the angle ($\Delta\theta$) between magnetic moments can easily be obtained by minimizing equation (3.3):

$$\Delta\theta \approx \cos^{-1}\left(-J_L/2J_Q\right). \tag{3.4}$$

Therefore, depending on the coupling coefficients, the following cases have been discussed:

i. If, both J_L and J_Q are negative and:
   a. $|J_L| > |J_Q|$, the energy per unit area becomes minimum for AFM coupling ($\Delta\theta = 180°$).
   b. $|J_L| < |J_Q|$, the energy per unit area becomes minimum for a non-collinear spin structure with $\Delta\theta = 90°$.

ii. If J_L is positive, the energy per unit area becomes minimum for FM coupling ($\Delta\theta = 0°$).

iii. If J_Q is positive, the energy per unit area becomes minimum for both FM and AFM coupling.

Therefore, a large negative value of J_Q induces the non-collinear spin structure. In addition to the study by Rührig *et al* [57] on Fe/Cr/Fe heterostructures, many other groups studied the biquadratic interaction in several FM/NM/FM heterostructures [58–64] during the same time.

Initially, the interlayer coupling was mostly measured indirectly by measuring the magnetic properties such as the hysteresis loop using macroscopic techniques such as MOKE [65], and a superconducting quantum interference device (SQUID). The exchange coupling is inferred by comparing the measured property (hysteresis curve) with a model. There is an imaging technique that has also contributed significantly to understanding interlayer exchange coupling, called scanning electron microscopy with polarization analysis (SEMPA) [63, 64]. This technique is based on measuring secondary electrons. It is generally not used with an applied field, limiting it to studies of the remnant state. The other indirect techniques used to measure the coupling are based on determining magnetization direction-dependent energy curvature. Two such techniques are ferromagnetic resonance (FMR) [66] and Brillouin light scattering (BLS) [49, 67]. In a multilayer structure, certain resonances depend on the strength of the interlayer coupling, whether as a function of the field at a fixed frequency (FMR) or frequency at the fixed field (BLS) and thus the resonance positions can be used to determine the coupling [66, 67]. As mentioned above, the nature of the exchange coupling and coupling strengths are usually determined from magnetic hysteresis measurements using MOKE or SQUID. However, a non-collinear coupling is often encountered in exchange-coupled superlattices and hysteresis measurements do not properly reveal sufficient information for understanding the coupling such as (i) the angle between the magnetization vectors in the adjacent ferromagnetic layer, (ii) the extent of the magnetic correlation perpendicular to the ferromagnetic layer, and (iii) the distribution of in-plane magnetic domains. Moreover, the residual antiferromagnetic coupling is often weak and can usually not be distinguished from soft ferromagnetic behaviour. Thus, using intrinsic properties of spin-dependent PNR, such as the depth-sensitivity of subnanometer resolution and possessing vector magnetometric capabilities, one can reveal coupling angles between alternate FM layers, in-plane and out-of-plane correlation lengths, weak antiferromagnetic exchange coupling strength, domain size distributions, and spin disorder at the interface. In addition to the structural properties of magnetic/NM heterostructures, the magnetic interface roughness and/or the evolution of the roughness into the multilayer stack, the interlayer coupling and GMR properties also depend on the details of the magnetic domain structure [68, 69]. Interface roughness in these FM/NM/FM heterostructures is also believed to contribute to influencing the exchange coupling [70], and thus PNR plays an important role to investigate such effect experimentally. Neutron techniques have also been demonstrated in determining field dependence of the antiferromagnetically coupled domain structure [68, 71].

The most commonly studied AFM-coupled metallic multilayered systems are Fe/Cr and Co/Cu. The Co/Cu superlattice was one of the first systems in which AFM exchange coupling was successfully demonstrated in 1989 using polarized neutron scattering [72]. Later, in 1993, Schreyer *et al* [73] confirmed the oscillatory nature of the coupling behaviour in Co/Cu multilayers using PNR. In 1999 Borchers *et al* [74] studied a relatively thicker Co/Cu multilayer for understanding the influence of interface roughness on interlayer coupling in this system using PNR and found the in-plane domains of the Co layer are antiferromagnetically coupled across the Cu layer. Later, in the year 2000, using specular and off-specular PNR, Langridge *et al* [68] revealed a close relationship between domain disorder, domain size, interlayer coupling, and the GMR effect in Co/Cu multilayers.

For the first time, the ordering of magnetic moments with an intermediate angle (~50°) instead of bilinear or biquadratic has been observed experimentally for $[Fe/Cr]_N$ superlattices by Schreyer *et al* [75] in 1995 using PNR and Kerr magnetometry. Figure 3.6 shows the spin-dependent PNR data with polarization analysis (non-spin-flip and spin-flip) and analysis for an Fe/Cr multilayer of nominal

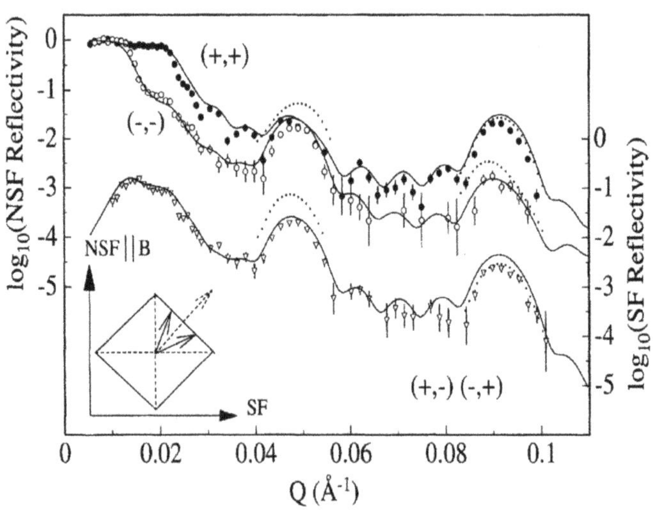

Figure 3.6. Non-spin-flip (NSF: +,+ and −,−) and spin-flip (SF: +,− and −,+) PNR data (symbols) and corresponding fits assuming a coupling angle of 90° (dotted lines) and 50° (solid lines) between two Fe layers in Fe/Cr multilayer with a structure of GaAs/Fe/Ag/[Fe (52 Å)/Cr (17 Å)]$_9$, where 9 is the number of bilayers. Reproduced with permission from [75]. Copyright 1995 the American Physical Society.

structure: GaAs (substrate)/Fe (10 Å)/Ag (1000 Å)/[Fe (52 Å)/Cr (17 Å)]$_9$, where 9 is the number of bilayers. The PNR data indicate a coupling of 50° (the fit with solid lines in figure 3.6) instead of 90° (the fitted curve shown by the dotted line in figure 3.6) between two Fe layers separated by the Cr layer of thickness 17 Å. Later Ustinov *et al* [76] found many coupling angles of 50°, 70°, 80°, 126°, and 143° in a series of Fe/Cr/Fe superlattices. Later, in 2002, Lauter-Pasyuk *et al* [77] demonstrated that the magnetization of the alternating Fe layers is twisted through the multilayer stack proving a stable non-collinear configuration. Recently, the nuclear resonance reflectivity (NRR) technique of a synchrotron source was also used to investigate the depth profile of magnetization where it uses a nuclear hyperfine method in which x-ray scattering is used after tuning the energy of the x-rays to the nuclear transition energy of a Mossbauer active element (e.g. ^{57}Fe) [78]. In this case, the total scattering amplitude from the Mossbauer isotope increases significantly due to the additional contribution from nuclear resonance scattering [78].

Using NRR Andreeva *et al* [79] observed a double-spiral magnetic structure of the [^{57}Fe/Cr]$_{30}$ multilayers. They found that two magnetic spirals, relating to the odd and even Fe layers, with different signs of rotation for multilayer with the intermediate thickness of the Cr spacer layer between the FM and AFM IEC. The IEC is also observed in many more magnetic systems, e.g. Fe/MgO superlattices [80], Fe/Si/Fe heterostructures [81], and others with impurity-mediated exchange coupling in Heusler alloys [82], dilute magnetic semiconductors [83], and CuMn/Co multilayers [84]. The temperature and field dependence of IEC in these systems are mostly investigated by probing the depth-dependent magnetic structure in the heterostructures using PNR. Saerbeck *et al* [84] investigated the magnetic impurity-mediated IEC observed in $Cu_{0.94}Mn_{0.06}$/Co multilayers using PNR and obtained a variation in the coupling angle of Co moments with respect to temperature and magnetic field. These examples indicate that the spin-dependent PNR with polarization analysis has been an extremely powerful technique for the quantitative determination of coupling angles in non-collinearly coupled superlattices and there is no other experimental technique that can provide such detailed information.

3.4 Multilayer with noncolinear and chiral magnetic structures

Both FM and AFM order tends to favour the collinear arrangement of the spins due to their energetic favouring of parallel and antiparallel ordering, respectively. Thus, for obtaining non-collinear spin textures including helical, conical, and chiral magnetic order, a fundamental difference in the underlying magnetic ground state is essential. While fundamental magnetic excitations (spin waves) disturb the linear ordering of the ground state, the stabilization of a non-collinear state is only found when there is some form of frustration/disorder in the spin interactions. Frustration is the inability for all interactions to be energetically minimized simultaneously and it generally occurs from a balance of all interactions due to the underlying geometry or the strengths of competing interactions. In bulk crystals a particular geometrically frustrated non-collinear AFM order with chirality has been stabilized in the ground state and the chirality is determined by the DMI, which arises due to a crystal structure with broken inversion symmetry [85]. The material choice of bulk chiral magnets with a broken inversion symmetry is still limited. In magnetic multilayer systems, many emergent phenomena, such as perpendicular magnetic anisotropy (PMA) [86], DMI [87], the Rashba effect [88], spin-Hall effect (SHE) [89, 90], spin-transfer torque (STT) [91], spin–orbit torque (SOT) [92], etc, have been studied for both fundamental research and applied perspectives. All these phenomena are closely associated with spin–orbit coupling (SOC). We will briefly discuss some of these phenomena in this section. The SOC is also important to understand other related phenomena including magnetocrystalline anisotropy, anisotropic MR, magnetic damping, and the anomalous Hall effect.

The SOC is the relativistic interaction between electrons' spin and momentum degrees of freedom and it occurs whenever a particle with non-zero spin moves around a region with a finite electric field (figure 3.7(a)). It can be written as

$$H_{\mathrm{SO}} = \frac{Ze^2}{2m_e^2c^2r^3}\sigma.\ L, \tag{3.5}$$

where σ, L, Z, m_e, e, c, and r, are the electron spin angular momentum, the orbital angular momentum, the atomic number of the nucleus, the mass of the electron, the electronic charge, the speed of light, and the radius of the orbit, respectively. Considering the expected value of $< \boldsymbol{r}^3 > \sim < \boldsymbol{Z}^3 >^{-1}$, the atomic SOC scales with Z^4, which explains why the SOC is strong in heavy elements such as Pt, Ta, and Ir. However, orbital hybridization also plays a significant role when considering complete crystals rather than single atoms. Thus, in ultra-thin film multilayer systems a noticeable interfacial DMI can arise due to the large SOC in the presence of broken inversion symmetry at the interfaces (figure 3.7(b)). DMI is an asymmetric exchange interaction and this phenomenological model was first proposed by Dzyaloshinskii [7] in 1958 to explain the weak ferromagnetic effect observed in antiferromagnetic crystals such as α-Fe_2O_3. Later Moriya explained this theoretically with SOC in 1960 [8]. In 1980 Fert and Levy discovered large DMI between Mn spins in the presence of nonmagnetic impurities such as Pt or Au [93]. The energy associated with DMI between two neighbouring atomic spins $\boldsymbol{s}_i$ and $\boldsymbol{s}_j$ can be written as (figure 3.7(b)):

$$E_{\mathrm{DMI}} = -D_{ij}\cdot\left(S_i \times S_j\right), \tag{3.6}$$

where D_{ij} is the DM vector, which is perpendicular to both the axis of broken symmetry and the vector connecting two spins. The presence of DMI in a magnetic system favours non-collinear spin alignment—canting the spin alignment and leading to chiral spin textures. When the DMI is non-negligible, a competition between symmetric (Heisenberg) exchange interaction and antisymmetric DMI emerges leading to the creation of non-collinear magnetic phases such as helical, conical, bubble, vortex, and the so-called skyrmion phases. A non-zero DMI has been achieved in artificial heterostructures of ultra-thin FM layers with heavy elements where there is broken inversion symmetry at the interfaces as well as the strength of the SOC being enhanced [94, 95].

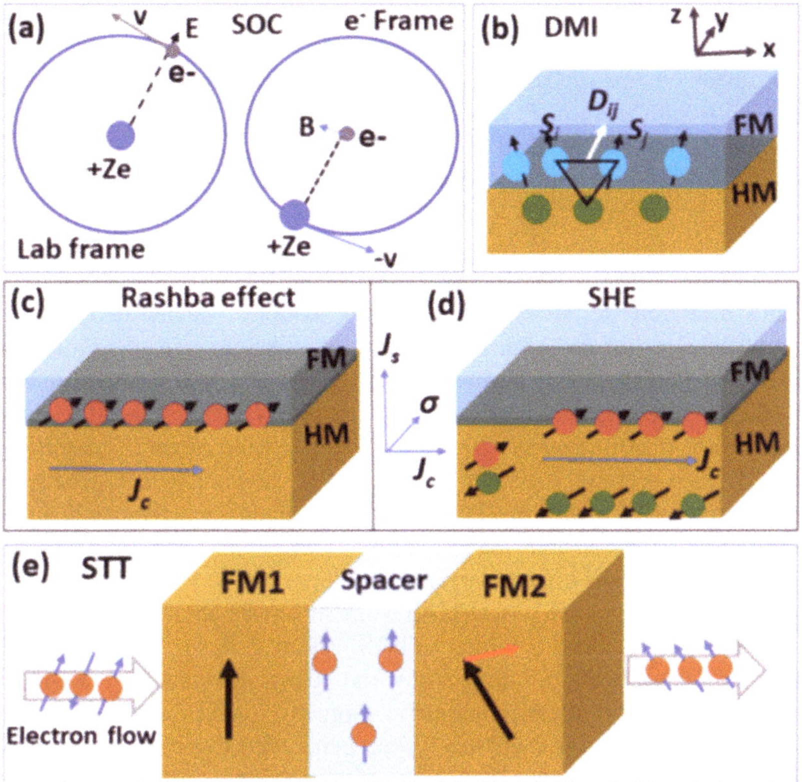

Figure 3.7. (a) Schematic showing an atom from the frame of reference of its nucleus, +Ze (left) and the electron (right). In the electron's frame of reference, an electron observes a positive nucleus of charge +Ze orbiting with velocity $-v$ and forming a closed current loop around it. Thus the electron experiences a magnetic field due to its orbital angular momentum and gives spin–orbit coupling (SOC). (b) Schematic of a bilayer exhibiting the interfacial Dzyaloshinskii–Moriya interaction (DMI) between the FM and the HM layers. The DMI is mediated via the large SOC of the HM atoms to the magnetic atoms located above them, which leads to a canting of the ferromagnetic spins. The DMI vector, $\boldsymbol{D}_{ij}$ points along the y-axis parallel to the plane of the interface. (c) The Rashba effect in a heterostructure with broken inversion symmetry. An internal electric field (E) is generated at the interface along the direction of symmetry breaking due to the potential drop and the conduction electrons with momentum P move near the interface. An effective magnetic field is induced and thus the SOC effect produces spin-polarized electrons at the interface. (d) Schematic of the spin-Hall effect (SHE) in a spin source material. When an unpolarized charge current (J_c) flows into a source material with SOC, electrons with the spin-up and -down deflect in the opposite direction due to the spin-dependent asymmetric scattering, leading to a transverse spin current (J_s) perpendicular to the direction of both the charge current and the polarization (σ) of the accumulated spins. (e) Schematic of the STT effect in one pinned layer (FM1) and one free layer (FM2) separated by a nonmagnetic layer (spacer) with an unpolarized current flowing through the multilayers. It suggests that the spin current can switch the magnetization direction of the free layer (FM2).

The electronic band structure can be modified by splitting degenerate bands with finite angular momentum (p, d, and f) as a result of SOC. The Rashba effect and SHE are the two most general ways to generate pure spin current in source materials with large SOC using electrical methods. The SOC of the spin of itinerant electrons with the magnetic field generated by the resultant electric field due to the uncompensated bonds formed at the surface or interface because of broken inversion symmetry is known as the Rashba effect [88]. This effect produces spin-split dispersion even on the surfaces of conventional (nonmagnetic) metals. Electrons travelling with momentum P at the interface will experience a magnetic field perpendicular to the electric field (E_0) along the plane perpendicular to the film (z-direction). The Rashba Hamiltonian can be written as

$$H_R = \alpha_R(\boldsymbol{\sigma} \times \boldsymbol{P}) \cdot \hat{z}, \tag{3.7}$$

where $\alpha_R = \gamma\mu_B E_0/2mc^2$ is the Rashba constant, and σ and z are the Pauli-matrix vector and direction of the electric field. H_R couples the momentum and spin direction of the electron and gives rise to a non-equilibrium spin accumulation at the interfaces. The Rashba effect (figure 3.7(c)) plays a key role both in fundamental physics and for potential applications. There has been considerable effort devoted to exploring and/or maximizing the Rashba effect in a wide range of materials [96].

The SHE has become a widely used strategy nowadays for generating and controlling spin currents without the use of a ferromagnetic reference layer [90, 97]. It is the phenomenon of spin accumulations along the edges of a nonmagnetic conductor while passing a charge current. In comparison to the Hall effect (HE) and anomalous Hall effect (AHE), the SHE results in a pure spin current and does not need a broken time-reversal symmetry (TRS) (figure 3.7(d)). While HE and AHE deal with the charge accumulation along the edges and need either the external magnetic field (HE) or internal magnetic moments (AHE) to break TRS, SHE defines the generation of a spin current (j_s) perpendicular to a charge current (j_c) and the spin-polarization direction σ as follows (figure 3.7(d)):

$$j_s = \frac{\hbar}{2e}\theta_{SH}\, j_c \times \sigma, \tag{3.8}$$

where θ_{SH} is the spin-Hall angle and describes the conversion efficiency. The origin of the SHE in heavy NMs, e.g. Pt, Ta, and W, etc, is rooted in the SOC of the material.

Theoretically, Slonczewski [98] and Berger [99] first calculated and revealed that the spin can transfer its angular momentum to the magnetic moment and exert torque on the FM layer's magnetic moment. This torque can switch the magnetization direction of the FM layer and it is called the spin-transfer torque (STT) effect. To explore the STT effect, the first step is to produce a spin-polarized current (figure 3.7(e)). The spin-polarized electron can be obtained by sending an unpolarized current through a magnetic layer (FM1) with a fixed magnetization direction. If the magnetization direction in the thin free magnetic layer (FM2) is different from the polarization direction of the spin, there is an STT effect acting on the magnetization of the free layer which rotates it. The magnetization direction of the FM2 layer can be reversed if the generated spin-polarized current density is high enough. Before the discovery of STT effect, the magnetization direction of a magnetic layer was changed by applying a magnetic field, but now it is possible to change the magnetization direction of the FM layer by sending spin-polarized current. Thus STT provides an alternative way to write data in magnetic random-access memory

(MRAM). When an in-plane current is applied to an NM/FM heterostructure, a spin-polarized current, generated in the NM, flows to the FM and exerts torque. This torque, known as the SOT, can manipulate the magnetization of the FM layer. Compared to the classic STT, SOT exhibits superior performance in terms of efficiency and speed. Consequently, the SOT technique becomes very promising for practical applications and may even reshape the landscape of spintronics.

It is evident from the various effects described above that the need to understand magnetic moment density in various layers in a heterostructure and the determination of the magnetization vector in the layers is of utmost importance for understanding the fundamental mechanism and its application towards designing magnetic devices. In this respect, the intrinsic properties of PNR have become an inevitable tool for characterizing a heterostructure with excellent spatial resolution.

3.4.1 Transition metal/transition metal multilayers

Transition metals (TMs) are categorized along three series in periodic table: the 3*d* series from Sc ([Ar]$3d^1 4s^2$) to Cu ([Ar]$3d^{10}4s^1$) and Zn ([Ar]$3d^{10}4s^2$); the 4*d* series from Y ([Kr]$4d^1 5s^2$) to Ag ([Kr]$4d^{10}5s^1$) and Cd ([Kr] $4d^{10}5s^2$); and the 5*d* series from La ([Xe]$5d^1 6s^2$) to Au ([Xe]$5d^{10}6s^1$) and Hg ([Xe]$5d^{10}6s^2$). Transition metals display multiple oxidation states because of the proximity in the energy of the $(1 - n)$th *d* and *n*th *s* shells that make their chemistry particularly interesting, leading to a wide variety of transition metal oxides and organometallic compounds. The magnetism in isolated transition metal ions is governed by Hund's rules, on the other hand, the magnetic properties of bulk transition metals are determined by band magnetism and the Stoner criterion, which suggests that magnetism emerges as a compromise between the kinetic energy and exchange interaction. As we know that spin–orbit coupling increases with Z, and the 5*d* transition metals possess a much larger spin–orbit coupling than their 4*d* and 3*d* counterparts, heterostructures of FM 3*d* (Fe, Co, and Ni) metals with 4*d* (Zr, Mo, Pd, Rh, etc) and 5*d* (Hf, Ta, Ir, Pt, etc) elements are quite appealing for spin-orbitronics.

Proximity-induced magnetization is an emerging interfacial phenomenon observed in FM/heavy metal (HM) heterostructures, particularly in Fe/Pd heterostructures where the proximity of Fe induces a large moment in the Pd which extends ~2 nm from the interfaces [100]. While the interface between an HM element with strong SOC and an FM transition metal develops different magnetic spin structures due to interfacial DMI, the systems with both DMI and proximity-induced magnetization may make highly promising candidates to realize high-speed and energy-efficient memory devices [101]. Using *in situ* PNR, Mayr *et al* [101] observed proximity-induced magnetism as well as a Dzyaloshinskii–Moriya interaction located directly at the Pd/Fe interface of a polycrystalline Pd(110 Å)/Fe(4.1 Å)/Pd (720 Å) trilayer heterostructure. This observation was in line with that of the Co/Pt system [102], however, no correlation between the presence of proximity-induced magnetism and DMI was reported in this system.

Perpendicular magnetic anisotropy (PMA) is one of the important properties exploited in magnetic device applications and is mostly exhibited by FM (TM)/HM

(TM) heterostructures such as Co/Pt, Co/Pd, Fe/Pt, and FePt/Fe, etc, these multilayers have attracted a great deal of interest from the point of view of technological advances in high areal density magnetic recording [103].

3.4.2 Magnetic phase transitions in rare-earth/transition metal multilayer

Magnetic multilayers, composed of a rare-earth (RE) element, such as Gd, Tb, Er, or Dy, and a transition metal (TM), such as Fe, Co, FeNi, and Ni, are exciting systems which show different magnetic phases as a function of temperature and magnetic field due to their different ordering temperatures and strong AFM coupling at the interfaces. Early research activities on RE/TM heterostructures covered the investigation of magnetic exchange phenomena between TM and heavy RE layers such as exchange bias [104] and interfacial coupling [105]. The magnetic state in these heterostructures is determined by competition between the Zeeman energy and the interfacial exchange energy. Different temperature dependencies of RE and TM magnetization exhibit in-plane cancellation of moments at the compensation temperature (T_{comp}), where RE and TM moments are equal and opposite to each other, so the net moment in the system tends to zero [106]. The compensated RE and TM heterostructures, where the RE and TM moments are aligned antiparallel due to strong AFM interaction and the total net moment tends to zero at T_{comp}, are potential candidate materials for realizing devices with higher speed and density [107, 108]. Recent studies on a class of artificial ferrimagnets consisting of RE–TM alloys and heterostructures have also indicated the potential of these systems to exhibit domain wall (DW) motion via exchange coupling torque [109]. The use of nano-sized stable magnetic helices for magnetic energy storage applications has been predicted theoretically [110] and realized experimentally in RE/TM multilayers [111, 112]. It has been recognized that properties such as magnetization reversal and formation of helical configurations (in the form of planar 2π DWs) in RE/TM heterostructures without the application of magnetic field at the compensation temperature are key to manipulating the magnetic devices. In addition, the magnetic interaction at interfaces in these RE/FM heterostructures can be modified by exploiting interface morphology, and this can subsequently be utilized to engineer materials with magnetic properties that are significantly different from the bulk. Recently, the engineered magnetic multilayers of RE and TM were used as a key components of superconducting spin valves [113, 114] for possible applications in superconducting spintronics. Here we discuss emerging phenomena at the interfaces of different RE/TM heterostructures.

3.4.2.1 Gd/TM-based heterostructures

In recent decades much attention has been devoted to ferrimagnetic thin films consisting of TM/RE heterostructures and their intermetallic alloys [115]. The Gd/TM-based ferrimagnetic systems, in particular, Fe/Gd multilayers, have been studied intensively. Gd is the only near room temperature (RT) ferromagnet, with a Curie temperature (T_c) of ~293 K, among the RE materials. The permanent magnetic moment of Gd is 7.98 μ_B, which is much more than that of TM atoms in their bulk, e.g. Fe (2.2 μ_B), Ni (0.6 μ_B), and Co (1.8 μ_B). Different temperature dependencies of Fe and Gd magnetizations and strong AFM coupling at their interface can lead to a T_{comp}, forming an artificial antiferromagnet. They are also an excellent demonstration for a variety of collinear and twisted magnetic states [116–119]. Beyond T_{comp} the magnetic phase diagram consists of aligned phases at low fields (exchange-dominated) and twisted phases at high fields (exchange-Zeeman competition). Different experimental techniques are used to explore the twisted states of spins in Fe/Gd multilayers, such as magnetometry, magnetoresistivity, MOKE, PNR, x-ray resonant magnetic scattering (XRMS), and x-ray magnetic circular dichroism (XMCD) measurements [105, 118–123]. Recently, Gd/Fe-based multilayer systems have also been studied for various topologically nontrivial spin objects such as skyrmions and antiskyrmions [124, 125]. Using x-ray resonant scattering, Haskel *et al* [105] found enhancement of the Gd-ordered magnetic moment near Gd/Fe interfaces. Combining PNR and x-ray resonant magnetic reflectivity (XRMR) techniques and a simultaneous refinement of these data from [Fe (35 Å)/Gd (50 Å)]$_5$ multilayers, Kravtsov *et al* [119] determine the magnetic structure as a function of temperature and magnetic field. They showed that a small number of bilayer periods together with an asymmetric termination (Fe-top, Gd-bottom) lead to unique low-temperature magnetic phases with twisting of Fe and Gd magnetic moments and nonuniform distribution of vectorial magnetization within Gd layers. A twisted magnetic state was found to be stable at small magnetic fields and at a low temperature of 20 K, which is well below the T_{comp} of this artificial ferrimagnetic system.

The Gd/Co multilayer system is another artificial ferrimagnet, which has been investigated for many static and dynamic magnetic configurations. Using mean-field calculations, Camley *et al* [106, 126] showed the existence of several magnetic states for Gd/Co and Gd/Fe multilayers. It was suggested that the twisted (or helical) magnetic state with the non-collinear configuration of magnetization over a film's thickness can appear at the T_{comp}. However, it is difficult to measure detailed magnetic structure across the interfaces of RE/TM heterostructure using macroscopic magnetometry techniques due to the interface-driven nature of RE/TM systems. The non-collinear (helix) magnetic structure can be studied using depth-dependent magnetic characterization techniques, e.g. nuclear resonant scattering [127], resonant magnetic x-ray reflectometry [128], or PNR [112], and they have their advantages and disadvantages. The helical structure of ferromagnetic alignment in heterostructures can be distinguished easily using spin-dependent PNR with polarization analysis. Singh *et al* [112] found a twisted helical magnetic structure with a planar 2π domain wall structure (DWs) and highly correlated magnetic domains in a Gd/Co multilayer near T_{comp} using spin-dependent specular and off-specular PNR. Basha *et al* [129, 130] studied the correlation of magnetic and magnetotransport properties across T_{comp} of Gd/Co multilayers grown with different interface morphologies. Interfacial dependence was achieved by varying the deposition conditions as well as by annealing the multilayer samples.

A comparison of magnetic and magnetotransport properties across T_{comp} of Gd/Co multilayers grown under different Ar pressures are shown in figures 3.8 and 3.9.

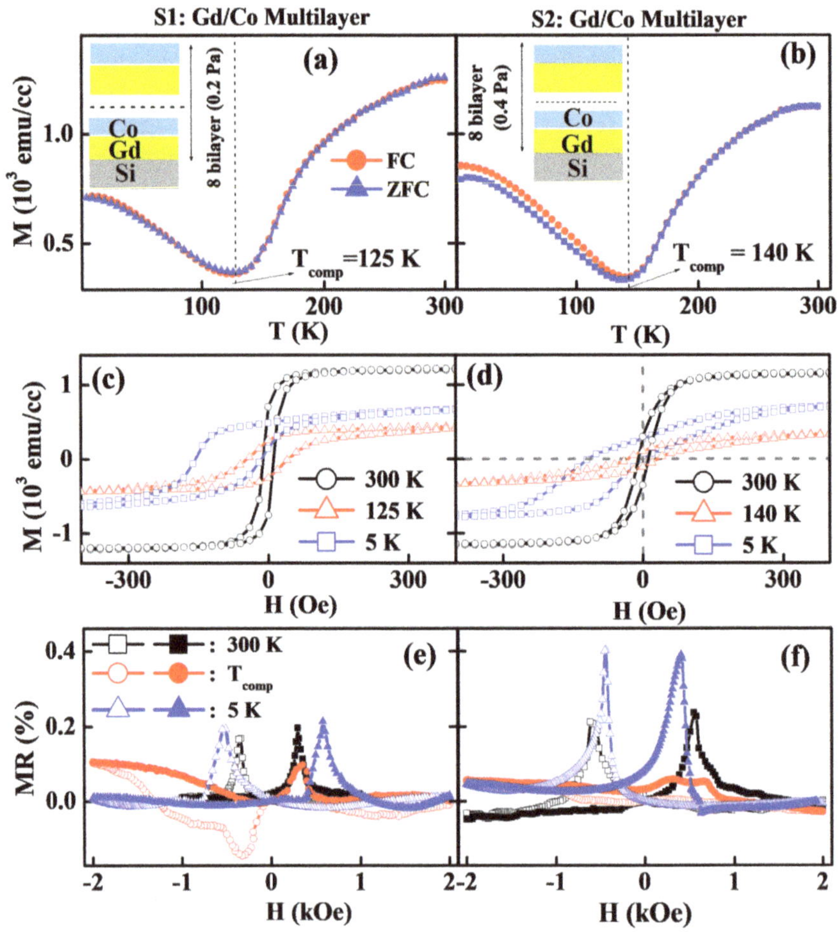

Figure 3.8. Magnetization data as a function of temperature ($M(T)$) under FC (in-plane applied field of 500 Oe) and ZFC conditions for multilayers (a) S1 and (b) S2. The insets of (a) and (b) show the schematic of Gd/Co multilayers S1 and S2 with eight bilayer periods each and grown under an Ar pressure of 0.2 Pa (S1) and 0.4 Pa (S2). DC magnetization ($M(H)$) curves at different temperatures for multilayers (c) S1 and (d) S2. MR (%) data in an applied in-plane magnetic field (longitudinal direction: current and field are applied in the same direction) at different temperatures for the multilayers (e) S1, (f) S2. Reproduced with permission from [112]. Copyright 2019 the American Physical Society.

Two Gd/Co multilayers (samples S1 and S2) of similar thicknesses were grown on Si substrates under different argon pressures of 0.2 and 0.4 Pa, respectively, and the schematics of these multilayers are shown in the insets of figures 3.8(a) and (b). Temperature-dependent magnetization ($M(T)$) of S1 and S2 multilayers in FC and ZFC modes are shown in figures 3.8(a) and (b). The FC data were recorded under the application of an in-plane magnetic field of 500 Oe. The dip in the $M(T)$ data where they show the minimum of magnetization (signature of antiferromagnetic coupling) is called T_{comp}, which is ~125 and ~140 K for S1, and S2, respectively.

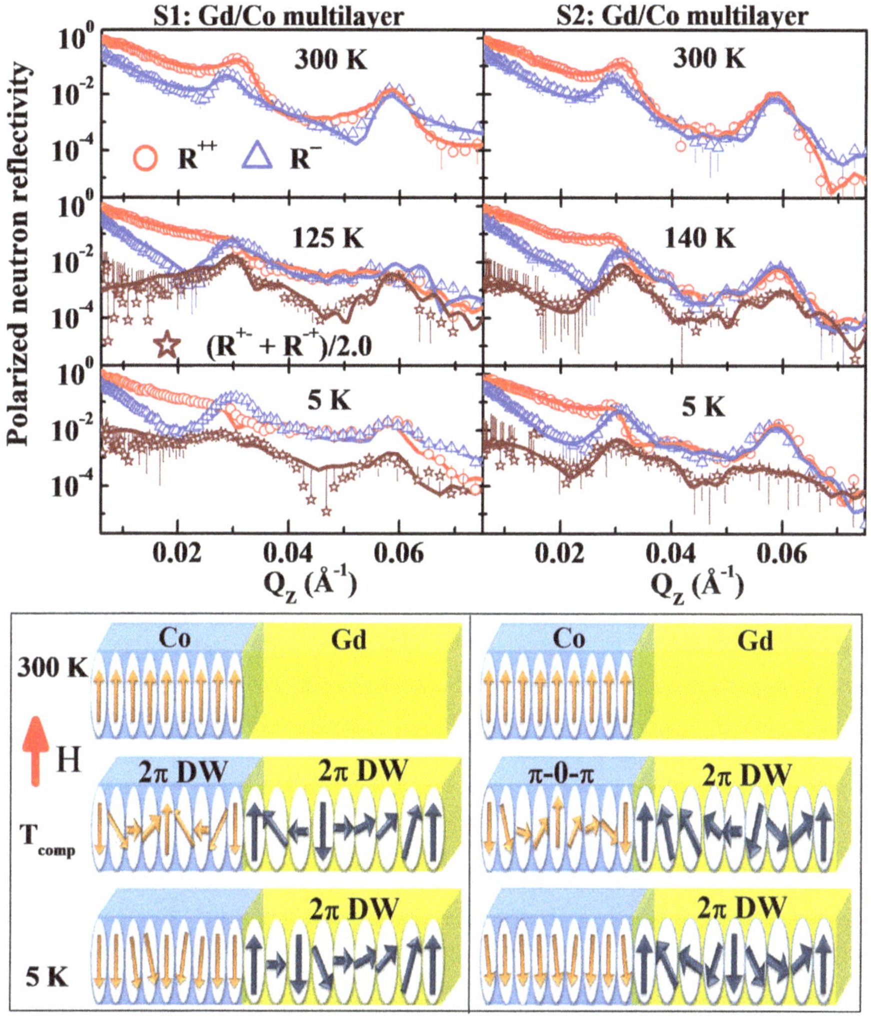

Figure 3.9. Spin-dependent specular PNR data (NSF: R^{++} (red open circles) and R^{--} (blue open triangles); SF: $(R^{+-} + R^{-+})/2.0$ (maroon star)) along with fit (solid lines) at different temperatures for the multilayers S1 (left panel) and S2 (right panel). Representation of magnetic structures along the depth of a bilayer of Gd/Co in multilayers S1 (left panel) and S2(right panel), at different temperatures obtained from PNR data. Reproduced with permission from [112]. Copyright 2019 the American Physical Society.

In addition, multilayer S2, which was grown at higher Ar pressure, showed a bifurcation between FC and ZFC below the T_{comp}, which may be due to different growth conditions affecting the interface morphology in these multilayers. The magnetization hysteresis loops $M(H)$ in a small field range (±500 Oe) and at different temperatures across T_{comp} for S1 and S2 are shown in figures 3.8(c) and (d), respectively. Interestingly, the temperature dependence of $M(H)$ for multilayers showed distinct behaviours with S2 showing asymmetric hysteresis loops and vertical shifts at low temperatures. Both the multilayers showed a shift (exchange bias) of $M(H)$ loops towards the negative field axis below T_{comp}. Magnetoresistance (MR) data were measured at different temperatures and are shown in figures 3.8(e) and (f). Interestingly we obtained a completely different MR behaviour for these multilayers at and around their T_{comp}. At T_{comp} (125 K and 140 K) for S1 and S2, an irreversible and antisymmetric MR (opposite variation of MR on sweeping the magnetic field in the two directions) was observed in both the multilayers; however, it was more prominent in the case of S1. The additional irreversibility in the field-dependent MR data for multilayers S1 and S2 near T_{comp} may be due to the depth-dependent magnetic structure within the Co and Gd layers of these multilayers.

To address this specific issue, the spin-dependent PNR measurements were carried out and results are shown in figure 3.9. Specular PNR with polarization analysis, i.e. non-spin-flip (NSF), R^{++} and R^{--}, and spin-flip (SF), R^{+-} and R^{-+}, reflectivities are used to determine the magnitude and direction of the magnetization vector along the depth of the multilayers. Figure 3.9 shows the specular PNR data as the NSF (R^{++} (red open circles), R^{--} (blue open triangles)) and SF (($R^{+-} + R^{-+}$)/2.0 (maroon star)) reflectivities and corresponding fits (solid lines) as a function of the wave-vector transfer Q_z, normal to the sample surface, at different temperatures for S1 (left panel) and S2 (right panel). The specular reflectivity data are collected up to a Q_z of ~0.08 Å^{-1}, which includes two Bragg peaks due to the periodic repetition of bilayers at Q_z ~ 0.03 Å^{-1} (first order) and 0.06 Å^{-1} (second order) indicating a bilayer thickness of ~214 Å. At 300 K no SF signal was observed in both the multilayers, suggesting that the Co moments in the different layers are aligned along the applied field (the Gd layer has zero moments at 300 K, as T_c for the Gd is ~293 K), as represented in the bottom panel of figure 3.9. At lower temperatures strong SF reflectivity was obtained, suggesting the existence of a perpendicular (in-plane) component of the magnetization in both multilayers. Attempts to fit the PNR data for both the multilayers, where SF reflectivity was observed, with homogeneous Gd and Co layers failed to reproduce the observed results and thus a helical magnetic structure for each layer was necessary to fit the data. The individual Co and Gd layers were split into sub-layers with a constant magnetic moment within the Co and Gd layers but a varying angle of rotation of the magnetization with respect to the applied field, i.e. a helical structure. The magnetic structure for a bilayer of the multilayers S1 and S2 at different temperatures, obtained from PNR data, are shown in the lower panel of figure 3.9. PNR data at T_{comp} for S1 suggested a 2π rotation of moments in both the Gd and Co layers and form a planar 2π DW structure. In contrast, for S2 we obtained a 2π rotation of moments for both the Gd and Co layers with a full 2π (0 to 2π) rotation of moments in the Gd layer and a π–0–π rotation of moments in the Co layer at T_{comp} (figure 3.9 (lower panel)). At 5 K both samples showed similar behaviour: Co layers FM and Gd layers 2π DWs. Thus specular PNR provided a detailed depth-dependent magnetic structure of multilayers at different temperatures and suggested the formation of planar 2π DW, within both the Co and Gd layers of S1 and S2 at T_{comp}.

Using the element-specific measurement technique of x-ray magnetic circular dichroism, Higgs *et al* [131] studied Gd/Ni heterostructures and determined the full magnetic state diagrams as a function of temperature and magnetic layer thickness.

A modified Stoner–Wohlfarth-based model was used to compare the experimental data and they provided evidence of a thickness-dependent transition to a magnetic fan state which is critical in understanding magnetoresistance effects in RE/TM systems. Lapa *et al* [132] studied the temperature and magnetic field behaviour of magnetization in Gd/$Ni_{0.81}Fe_{0.19}$ (Py) heterostructures, using a combination of magnetometry and magnetotransport techniques. They determined the angle between magnetizations in the Gd/Py heterostructure and the magnetic field at different temperatures, by measuring the angular dependence of resistance.

3.4.2.2 Tb/TM-based heterostructures

The Tb/Fe multilayer system has been studied extensively for different magnetic phases and related applications. The Tb in bulk shows several phase transitions with a paramagnetic (PM) to the AF phase at around 227 K and exhibits a helical-AF structure between 229 and 221 K. The FM phase in bulk Tb is observed below 213 K. The high spin–orbit coupling of Tb renders Tb/TM a promising material, exhibiting larger magnetocrystalline anisotropy than Gd-based heterostructures. Tb/Fe multilayers were immensely investigated for PMA and large coercivity at low temperatures using ^{57}Fe Mossbauer spectroscopy, SQUID magnetometry, and polar MOKE [133]. The study suggested that the perpendicular Fe spin orientation and PMA in Tb/Fe multilayers originate from the interfaces. The interface-driven PMA in this system was further confirmed by another study where the interfacial properties of Tb/Fe multilayers were modified by swift heavy ion irradiation [134]. This study suggested that an increase in interface roughness in the Tb/Fe multilayer due to ion irradiation resulted in a decrease in PMA properties. In the year 2004 Takanod *et al* [135], using the XMCD technique, estimated the magnetic moment of both Tb and Fe as a function of field and temperature. They found a twisted magnetic structure at low temperatures, which was attributed to the strong competition among the exchange coupling, the Zeeman energy, and the anisotropic energy. Paul *et al* [136] studied field- and temperature-dependent magnetic properties of the [Fe (30 Å)/Tb (60 Å)]$_5$/Fe (45 Å) multilayer grown on a Si substrate and found a double hysteresis loop (DHL) with exchange-bias-like shifts along and opposite to the field cooling axis below the ordering temperature of the Tb. Using PNR measurements they confirmed an antiferromagnetic alignment between the individual layers of Fe and Tb and the formation of 2π-domain walls (DWs). The shifts of the hysteresis loops were attributed to the formation of 2π-DWs by the interface moments that are pinned by the magnetically hard Tb layers forming bidomain-like states in this layered artificial ferrimagnetic system. Later Mukherjee *et al* [137] observed a larger exchange bias in a [Tb (60 Å)/$Fe_{0.4}Cu_{0.6}$ (30 Å)]$_{20}$ multilayer grown on Si substrate which was attributed to additional pinning of moments within the individual clusters formed upon Cu dilution. They have also not found any DHL characteristics in the temperature-dependent $M(H)$ loops for this diluted Fe/Tb system. PNR data from this diluted heterostructure confirmed an antiferromagnetic coupling, which changes the sign of the individual layer magnetizations but remains strongly antiferromagnetically coupled up to the higher magnetic field. Therefore, the nonmagnetic dilution in the Tb/Fe heterostructure system can directly alter the formation of planar DWs.

The exchange coupling and helical phase characteristics in another RE/TM (Dy/Fe) multilayer were also investigated by Fust *et al* [111]. Magnetization measurements for Dy/Fe multilayers suggested an exchange bias field ~1.0 kOe at 10 K. While the PNR revealed the formation of twisted magnetic helices and direct evidence of helices in the form of planar 2π-DWs within both layers of Fe and Dy. The helices within the Fe layers were attributed to the reasonably strong induced in-plane magnetocrystalline anisotropy of Dy and the exchange coupling at the Fe–Dy interfaces.

3.4.2.3 RE/TM-based alloys

After the first observation of all-optical switching (AOS) of the magnetization of ferrimagnetic GdFeCo alloys [138], RE–TM alloys have become a material system of great interest. The ferrimagnetic RE–TM alloys exhibit a unique property: the compensation point, where the net moment of the alloy is zero, due to the magnetic moment of the RE element being equal in strength to that of the TM element because of the different temperature dependence of the magnetization of the two elements. This occurs only for certain combinations of temperature and composition, i.e. at a given temperature there will be a certain compensation composition, and a given composition will correspond to a certain compensation temperature. In addition, the coercivity diverges near the compensation point. Most of the investigated alloy materials and heterostructures show helicity-dependent AOS (HD-AOS) in response to circularly polarized laser pulses. The Gd-based RE–TM alloy led to a lower magnetic anisotropy in the alloy, which may pose problems for long-term information retention as well as the down-scaling of the memory bit size for magnetic device application. In this respect, the high spin–orbit coupling of Tb renders TbCo a promising material, exhibiting larger magnetocrystalline anisotropy than Gd-based alloys, which could be more relevant for HD-AOS spintronic-based applications. The TbCo alloys also exhibit a favourable PMA, which makes them attractive for densely packed magnetic memory elements. Binary and ternary RE–TM alloys such as GdFeCo, GdCo, TbCo, or GdFe have shown promising applications in the field of spintronic devices, including spin valves for magnetic read heads [139], perpendicular magnetic tunnel junctions (p-MTJs) [140], or the spin–orbit-torque phenomenon [141].

3.4.3 Rare-earth/rare-earth heterostructures

Heavy RE metals show a complex magnetic structure and magnetic phase transitions depending on the temperature and magnetic field and thus have been actively studied for many years. Multilayer (superlattices) containing RE metals are artificial magnetic materials whose magnetic properties often differ significantly from the properties of the individual RE metal layers. Heterostructures consisting of magnetic RE metal and a nonmagnetic metal Y are relatively well studied for the oscillating interlayer exchange interaction in Gd/Y superlattices [142], incommensurate spiral magnetic order [143], and chiral magnetic order induced by the magnetic field [144] in Dy/Y superlattices. PNR measurements on a magnetic-based

RE/RE (Dy/Gd) system suggested the formation of a fan magnetic order in the Dy layers in comparatively low magnetic fields (H = 100 Oe) at temperatures below 170 K, which propagates coherently throughout the superlattice [145]. Using PNR, Paul [146] studied Er/Tb multilayers and found long-range magnetic order despite their competing anisotropies along the out-of-plane (Er) and in-plane (Tb) directions. Off-specular PNR was used to probe the vertically correlated magnetic structures which suggested the existence of magnetic vortex-like domains with magnetic helical ordering within the Er layers. Later Ye *et al* [147] showed experimental evidence of a superimposed helical magnetic configurations within both Dy and Tb layers of Dy/Tb multilayers using the PNR technique.

3.5 Exchange bias: interface magnetization

Exchange-coupled multilayer systems with two different magnetic materials introduce a new anisotropy within this compound system. One such compound material is exchange-coupled ferromagnetic–antiferromagnetic multilayers, which under certain conditions display a shift of the hysteresis curve, called exchange bias (EB). The EB effect has been known since the year 1956 when Meiklejohn and Bean [9] discovered a unidirectional anisotropy in a Co (FM) and its natural oxide (CoO, an AFM) system on cooling it below the Néel temperature of CoO. The unidirectional anisotropy was observed in torque measurements under an external magnetic field. The angle dependence was found to be proportional to sin(θ) instead of an expected sin(2θ) for the uniaxial anisotropy of pure cobalt. The other important observation of this study was a shifted hysteresis curve as shown in figure 3.10. They found that upon cooling the sample in an external field from above to below the ordering temperature of the antiferromagnet (Néel's temperature of CoO ~291 K), the resulting hysteresis curve was shifted opposite to the external field along the field axis (solid curve in figure 3.10). The shift of the hysteresis curve away from its origin is called the exchange bias field (H_{EB}).

Even though the EB effect was discovered 65 years ago [9] its microscopic origin has long been under debate, and several competing models and theoretical approaches have been proposed to understand this phenomenon. A detailed overview of the experiments dealing with exchange bias systems was reviewed by Nogués and Schuller [10]. In the meantime, the EB effect has become an integral part of modern magnetism with implications for basic research and numerous device applications ranging from so-called spin valves to read heads in magnetic memory devices. An EB bilayer consists of two key elements, the FM layer and the AFM layer, which show rather different magnetic and structural properties. While the FM layer can be studied using macroscopic magnetization techniques such as MOKE, VSM, SQUID, and MFM, this is not the case for the magnetic interface and the AFM layer. The interface between the FM and AFM layer has low volume, which makes it difficult to separate its contribution from the FM layer and most importantly for understanding the EB effect identifying interfacial magnetic properties is essential. PNR and soft x-ray magnetic scattering are non-destructive techniques that can reveal some of the key magnetic properties of the interface.

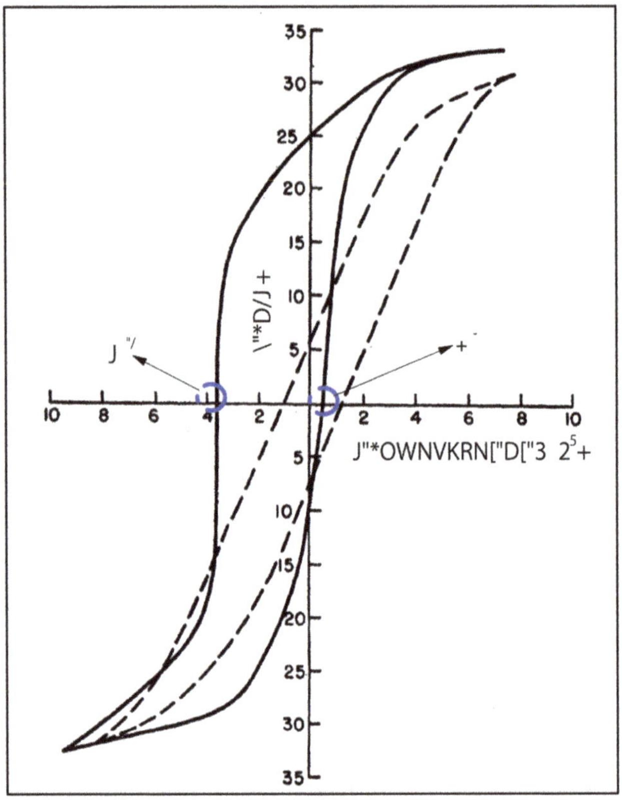

Figure 3.10. Hysteresis loop measured for Co/CoO system at $T = 77$ K after cooling with (solid curve) and without (dashed curve) applied magnetic field. Reproduced with permission from [9]. Copyright 1957 the American Physical Society.

Based on their observation, Meiklejohn and Bean [9] proposed a model to account for the magnitude of the hysteresis shift. They postulated a term for the total free energy including terms for the external field (H) which makes an angle θ with respect to the field cooling direction, the saturation magnetization of the FM (M_{FM}), its thickness (t_{FM}), its anisotropy constant (K_{FM}), the AFM layer's thickness and an interface coupling constant. Different energy terms can be added as needed to best account for the quantitative and qualitative behaviour of the macroscopic magnetization reversal. Hence, within this model, the energy per unit area, assuming coherent rotation of the magnetization can be written as

$$E = -\mu_0 H M_{\mathrm{FM}} t_{\mathrm{FM}} \cos(-\beta) + K_{\mathrm{FM}} t_{\mathrm{FM}} \sin^2(\beta) - J_e \cos(\beta), \tag{3.9}$$

where J_e is the interfacial exchange energy per unit area and β is the angle between the magnetization vector (M_{FM}) of the FM layer and the anisotropy direction of the

FM layer. The solutions of this equation are readily expressed in terms of the two coercive fields H^+ and H^- (see figure 3.10), which are the field values at which the hysteresis curve intersects the field axis corresponding to $\beta = 0$ and π. The connected coercivity (H_c) to the exchange bias field describes the width of the hysteresis loop, and it is defined as $H_c = (H^+ - H^-)/2$. The value of H_{EB} can be estimated using $H_{EB} = (H^+ + H^-)/2$.

During the last few decades, several models aiming to describe the experimentally found effects in EB systems were developed both qualitatively as well as quantitatively. The exchange bias is temperature-dependent and, generally, the EB field reduces with increasing temperature and vanishes at the so-called blocking temperature which may be much lower than the Néel temperature of the AFM layer. The EB effect also depends on the cooling field intensity, interface roughness, interface disorder, and crystallinity of the system [148, 149]. Sometimes, large cooling fields result in a reduction [150] or even a reversal of the EB [151]. The interface roughness is quite an important parameter, although in most models its influence on the EB is not fully understood. Typically, interfaces of AFM crystals with low roughness result in a small exchange bias, while rougher interfaces show an increased EB. Twinned antiferromagnets, containing two sub-lattices, e.g. Fe/FeF_2 and Co/CoO, may exhibit a higher exchange bias than fully epitaxial AFM layers [152]. In general, polycrystalline films often show a larger EB with smaller grain size.

Another interesting aspect of the EB phenomenon, known as the *training effect* is observed in FM/AF systems when they are field cycled. This effect is manifested in the reduction of the bias field and coercive field with consecutive hysteresis loops at a fixed temperature. It is generally believed that irreversible changes in the AFM layer result in this effect. Paul *et al* [153] showed that the training effect can be strongly influenced by the field cooling procedure for Co/CoO (FM/AFM) heterostructures and it was explained within the framework of the domain state model combined with a consideration of AFM grains in polycrystalline systems.

Experimentally, several materials and methods have been employed to investigate the EB phenomenon in magnetic bilayers and nanostructures. In general, the magnetic properties of these systems are highly affected by growth conditions and sample treatment, alloy composition, substrate temperature, film thickness, interface roughness and morphology, chemical stability of the alloys, and interdiffusion of atoms at the interface, etc. The fundamental mechanisms of exchange anisotropy have only recently been clarified experimentally. A quantitative predictive theory of EB is still elusive, mostly because the effect is fundamentally an extrinsic property of materials, dependent on defects, interface roughness, domain size, etc. The understanding so far is partial without the emergence of a unifying theory, which should involve explaining all different phenomena in different material systems with predictive capabilities. The lack of a unified understanding may be rooted in the fact that EB has continuously produced unexpected experimental 'surprises' which may agree with one model but not another. PNR play a crucial role to unravel the key issues and the hidden secrets of EB [112, 129, 154–162] because it is a tool that can directly and quantitatively determine the physical and magnetic structure over sub-nm depth resolution.

3.5.1 Exchange bias in FM/AFM core–shell systems

As mentioned earlier the EB effect had first been discovered at FM/AFM interfaces in fine Co/CoO nanoparticles [9]. The core–shell (CS) particles have both an interface and a surface, and so are prone to complex EB phenomena [163]. In general, the core is a highly ordered (bulk-like) magnetic object and the shell is just a ferrite layer with a widespread of local anisotropy axes [163]. Some of the largest observed EB effects have been reported for particles consisting of an FM core and an AFM (or ferrimagnetic) shell which has been grown around the core by chemical modification (usually partial oxidation) of the FM material. Both core and shell dimensions and their interaction, as well as the morphology of the interface, dictate the strength of the EB in such core–shell nanoparticles. However, this phenomenon can be better studied with PNR in tailored heterostructure with FM/AFM layers.

3.5.2 Exchange bias in transition metal (TM)/TM oxide heterostructures

Heterostructures consisting of TM FM and its oxide (AFM) such as Co/CoO, Ni/NiO, etc, have been studied extensively for the EB effect and related applications. Thin films of TM can be oxidized easily in air and thus form a TM/TM oxide thin-film system. Recently thin Co/CoO films deposited on an array of self-assembled polystyrene nanospheres were investigated and the EB effect was observed because of modification in shape anisotropy in this system [164]. The effect of AFM thickness and interface roughness is also studied as a perturbation in the interface magnetization, which further influences the EB effect in this system. Earlier PNR measurements were performed in Co/CoO multilayers to investigate the magnetization reversal and the time relaxation of the EB field close to the coercive field [165]. Ni/NiO is another system that may show interesting effects, although it is less often investigated than Co/CoO EB systems. Modification in EB effect in Ni/NiO heterostructures as a result of ion implantation [166], annealing [167], and strain [168], were also reported recently. A decrease in the EB field was observed on Cu ion implantation as well as on applying strain in Ni/NiO heterostructures. Fe, NiFe, NiFeCo, and CoFe are the other FM layers commonly used for a TM-based FM/AFM EB system due to their soft magnetic properties. The uniaxial and unidirectional anisotropies responsible for the EB effect in FM/AFM systems were studied in NiFe/NiO heterostructures as functions of the in-plane angle and it suggested a cosine dependent.

3.5.3 Exchange bias in other non-oxide-based heterostructures

The non-oxide-based systems such as FeMn, NiMn, PtMn, IrMn, FeF_2, MnF_2, etc, have been employed as an AFM layer in FM/AFM-based EB systems. This is because these materials may exhibit relatively simple crystalline structures with chemical stability and corrosion resistance, and are in general magnetically hard. In contrast to naïve expectations, the loop shift can sometimes be in the same direction as the cooling field ('positive EB'). This effect was discovered in FeF_2–Fe bilayers using high cooling fields, which originates from AFM coupling between the layers in contrast to the FM coupling assumed in most theoretical models. On the other hand, for the Ni/FeF_2 bilayer system, an unexpected precession of the magnetization in reverse magnetic fields that exceed the EB, originating from a reorientation of frustrated spins at the interface was observed. Artificially grown double-superlattices of Fe/Cr [169] and Co/Ru [170] have also shown the EB effect. While the EB in an Fe/Cr double-superlattice was determined by the thicknesses of the Cr spacer layer, where the EB field vanishes at a larger thickness of the spacer layer. The Co/Ru double-superlattice system exhibited a parallel domain wall, a spin flop transition, and EB when the anisotropy is large in the AFM block. A combination of PNR and MOKE measurements with energy minimization calculations suggested that neither the domain wall nor the spin flop is directly related to EB but that the anisotropy is essential. Using PNR measurements in both specular and off-specular mode, the dependence on the magnetic domain length scale as well as its distribution in Co/FeMn were correlated with the EB effect [171].

The PNR technique was also used to investigate the pinned and unpinned magnetization in a Co/FeF_2 system that was responsible for the EB in this system which was measured to be +167 Oe [154]. This study identified significant pinned magnetization in the AFM bulk and showed that unpinned Fe moments beneath the Co/FeF_2 interface were antiparallel to unpinned Co moments above the interface. Later the separation of the magnetization into pinned and unpinned components with nanometre depth resolution in the Co/FeF_2 system was investigated using PNR [155]. This system exhibited EB that could be switched from negative to positive just by cooling in weak or strong magnetic fields. The PNR measurements for Co/FeF_2 heterostructure were carried out in an applied field of ±7 kOe under the positive (EB field of +2.2 kOe) and negative (EB field of −2.2 kOe) EB conditions. The positive and negative EB were observed in heterostructure on cooling in an applied magnetic field of +7 kOe and 0.1 kOe, respectively [155]. Figure 3.11 shows the magnetization depth profile of the Al/Co/FeF_2 heterostructure obtained from PNR data for an applied field of +7 kOe (blue curve) and −7 kOe (red curve) for the sample with an EB field (H_E) of (a) −2.2 kOe and (b) +2.2 kOe.

The projection of the pinned magnetization depth profiles along (001) FeF_2 for $\pm H_E$ was estimated by determining the fractions of the magnetization depth profiles that do not change when the applied field is reversed and shown as dashed purple and orange curves in figure 3.11. The PNR results suggested that the distribution of pinned magnetization across the interface is nonuniform and is correlated with the sign of EB. The width of the interfacial magnetization extends between ~13 Å (for positive EB) and ~26 Å (for negative EB) away from the average chemical interface into the FM and AFM layers, which is larger than the width of the chemical interface.

Bilayers of soft and hard ferromagnetic thin films also showed remarkable EB. The EB effect was also observed in $Co_{0.66}Cr_{0.22}Pt_{0.12}$/Ni (hard/soft) ferromagnetic bilayers in a low applied magnetic field, whereas there was no shift in the hysteresis loop on applying larger field. The composite films consisting of $L1_0$ FePt as a hard magnetic phase and a soft magnetic (e.g. Fe, Co, etc) phase also exhibit EB. Basha *et al* [162] observed annealing-induced EB at room temperature in Fe–Cu–Pt-based

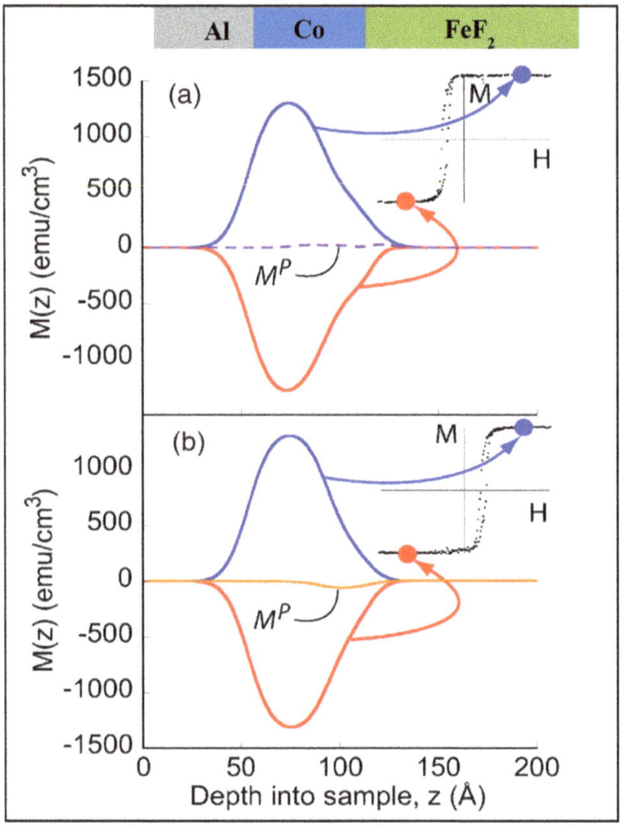

Figure 3.11. Magnetization depth profile of Al/Co/FeF_2 heterostructure obtained from PNR data for an applied field of +7 kOe (blue curve) and −7 kOe (red curve) for the sample with an exchange bias field (H_E) of (a) −2.2 kOe and (b) +2.2 kOe. The inset shows the schematic of the sample structure. The pinned components of the magnetization depth profiles are shown as dashed purple and orange curves. Reproduced with permission from [155]. Copyright 2007 the American Physical Society.

heterostructures with Cu as an intermediate layer (Fe/Cu/Pt heterostructure) and as a capping layer (Fe/Pt/Cu heterostructure), which was found to be dependent on the annealing temperature. A positive EB field of ~+120 Oe on annealing both the heterostructures at 400 °C, while on annealing these heterostructures at 500 °C and 600 °C a negative EB field of ~−100 Oe was observed and it was attributed to the coexistence of hard-soft magnetic phases along the thickness of the systems. Further, the observation of the annealing-induced EB effect in low-field regions is strongly dependent on the majority phase (hard/soft), present in the system.

3.5.4 Exchange bias in rare-earth/transition metal AFM-coupled heterostructures

Both positive and negative EB have been reported for RE/TM heterostructure systems [111, 112, 136, 172–174]. PNR measurements for an AFM-coupled GdFe/TbFe bilayer showed a direct correlation between the magnetic depth profile and the observed exchange bias [174]. The GdFe/TbFe bilayer that mimics antiferromagnetically coupled FM/AFM systems was studied over a large range of cooling fields and the frozen spin configuration in the pinning layer with changes in the magnetic structure as it undergoes training was observed. The results also showed a continuously varying EB field from negative to positive as the temperature increased, which was attributed to the relaxation, i.e. progressive untwisting, of a frozen interfacial domain wall located in the TbFe layer. Tb/Fe heterostructure studied earlier exhibited in-plane magnetic anisotropy and double hysteresis loops (DHLs) [111, 136], which was accompanied by EB-like shifts towards the negative and positive fields along the cooling axis. The $M(H)$ curve shifts are attributed to the formation of bidomain-like states in the AFM layers that are larger than the FM domains. PNR results suggested an induced magnetic moment for the Tb layer at RT (300 K) and the formation of 2π DWs within the Fe layers, which are responsible for the EB. Using spin-dependent specular and off-specular PNR, Singh *et al* [112] observed negative EB for Gd/Co multilayers (see figure 3.8) below T_{comp} and it was attributed to the formation of a twisted helical magnetic structure with planar 2π DWs and highly correlated magnetic domains in a Gd/Co multilayer near T_{comp}. Later it was found that these EB effects in the Gd/Co multilayer system were highly dependent on the interface morphology [130, 175].

3.5.5 Exchange bias in complex oxide heterostructures

Recently, a variety of complex oxide heterostructures have shown the EB effect, suggesting interface-dependent mechanisms of an altogether different means, e.g. charge transfer across interfaces, emerging phenomena of new magnetic coupling, etc, to achieve novel magnetic behaviour or even quantum mechanics. The EB in AFM insulator/paramagnetic metal [$CaMnO_3$(CMO)/$CaRuO_3$(CRO)] [161] and paramagnetic/ferromagnetic [$LaNiO_3$(LNO)/$La_{0.67}Sr_{0.33}MnO_3$(LSMO)] [160] heterostructures was observed at low temperature which was attributed to the development of interfacial FM order for CMO and LNO oxides at interfaces. PNR study suggested the formation of an FM interfacial layer in these systems which were coupled ferromagnetically and antiferromagnetically depending on the interface structure. $BiFeO_3$ (BFO)/LSMO is another complex oxide system that showed the EB effect at low temperatures [159]. BFO is a single-phase multiferroic material that exhibits magnetoelectric coupling between ferroelectric (T_c = 1103 K) and AFM (T_N = 643 K) order parameters. Observation of the EB effect in BFO/LSMO heterostructure was confirmed by PNR measurements and it was attributed to different interfacial coupling (FM or AFM) in the system due to possible surface termination and orbital reconstruction at the interfaces.

3.6 Proximity effect and coupling in complex oxide magnetic and superconducting heterostructures

The integration of FM and high-temperature superconductor (SC) is a central issue in developing superconductive spintronics [113, 176]. Investigation of the magnetic and superconductive properties in the hybrid heterostructure composed of FM and SC has been extensively done [177]. However, a prerequisite for the realization of such spin-polarized superconducting/ferromagnetic quantum states is the design and precise control of the magnetic order, in particular around the interface. With the conflict of magnetic compatibilities between both materials, an interface of the heterostructure is essential to accommodate their mutual interplay simultaneously. Many experimental studies have been reported for heterostructures composed of cuprate superconductor and colossal magnet oxides, i.e. YBCO/LCMO (or LSMO) [178–182] and the results play an important role in understanding carrier transfer, which deeply affect the proximity effect between the layers. YBCO in the bulk is an optimally hole-doped superconductor with a superconducting transition temperature (T_{SC}) = 90 K, and LCMO or LSMO, are half-metallic ferromagnets with bulk values of the Curie temperature of T_c = 270 and 330 K, respectively, and a saturation magnetic moment of 3.7 μ_B/Mn ion. It was found that a magnetic proximity effect (MPE) occurs at the YBCO/LCMO interface, which leads to a suppression of the ferromagnetic order on the manganite side that is commonly denoted as a magnetic 'dead layer'. The properties of YBCO/LCMO heterostructures are strongly influenced by a coupling phenomenon at the interface [177, 178], leading to complex behaviours in these heterostructures, namely magnetoresistance, transient photo-induced superconductivity, and MPE [176, 178, 179]. Competition between the two ordered ground states [177, 178] leads to the suppression of both superconducting and magnetic transition temperatures. A variety of exotic phenomena, which are thought to be closely related to the strong electronic correlations and the intimate coupling between the magnetic, orbital, and lattice degrees of freedom in these oxides, have been seen in these heterostructures. These yield a manifold of nearly degenerate ground states with a rich spectrum of properties that can be readily tuned with external parameters. Using PNR a giant modulation of the in-plane magnetization in LCMO layers was observed below the T_{SC} of YBCO/LCMO multilayers, which also depends on the sequence of deposition [179]. In another experimental study using PNR, depletion of magnetization or a magnetic 'dead' (MD) layer was found at the YBCO/LCMO interface and it was again demonstrated that length scales of the MD layer depend on the sequence of deposition of the YBCO/LCMO interfaces [176]. The result showed that the MPE-induced changes in the magnetic depth profile is governed by the electronic (orbital) state of the FM manganite layers and thus provided evidence for an intrinsic nature of the MPE and an intimate magnetic coupling across the cuprate–manganite interfaces.

Considerable efforts have been made to understand the magnetic modulation in the FM layer induced by superconductivity in these oxide-based FM/SC heterostructures, yet it remains unresolved. While some studies suggested that the orbital reconstruction and charge transfer play a role in the formation of the MD layer due to MPE, other studies indicate that the superconducting proximity effect below T_{SC} can be described in terms of leakage of the spin-triplet Cooper pairs into the FM leading to a lowering of the transition temperature in the SC. This means if FM–SC layers are interleaved with a band insulator, the leakage of Cooper pairs should stop and there will be an abrupt break in the long-range order of the triplet spin-paring. The best way to test the validity of this assumption is therefore to interleave the YBCO and LCMO layers with a layer of insulating (I) layer such as $SrTiO_3$ (STO).

In this case we should expect neither any induced magnetism of the Cu atoms on the YBCO side nor any reduction of the Mn magnetic moments on the LCMO side. In 2016 Prajapat *et al* [180] reported the influence of superconducting order parameters on an LCMO FM through an insulator STO barrier in YBCO/STO/LCMO epitaxial heterostructures using spin-dependent PNR. YBCO/STO/LCMO trilayer systems with varying thicknesses of STO layer were grown on an STO substrate and superconductivity-induced modulation in the magnetization depth profile in an FM layer was studied using PNR measurements across the T_{SC} [180]. Figures 3.12(a) and (b) show the normalized spin asymmetry (NSA), defined as $(R^{+} - R^{-})/R_{F}$, where $R^{\pm}$ are spin-dependent neutron reflectivity data normalized to R_{F}, $R_{F} = 16\pi^{2}/Q^{4}$ is the Fresnel reflectivity. The fits assumed different magnetization (M) depth profile in the YBCO/STO/LCMO hybrid system above (100 K) and below (50 K), respectively. The fitted magnetization depth profiles are shown in figures 3.12(c) and (d) for 100 and 50 K, respectively. The study suggested formation of an MD layer at the LCMO/STO (LCMO on STO) interface below T_{SC}. The continuous and dashed lines in figures 3.12(a) and (b) indicate the fit to PNR data using a model with and without the MD layer, respectively. From the fits to the data at 100 K and 50 K, it is evident that the magnetization model with the MD layer at the interface appears only at 50 K, below the T_{SC} of the sample. The results also showed that the length scale for the MD layer was significantly decided by the thickness of the insulator layer, with the thinner the insulator layer, the thicker was the MD layer at the LCMO/STO interface [180]. This was attributed to tunnelling of the

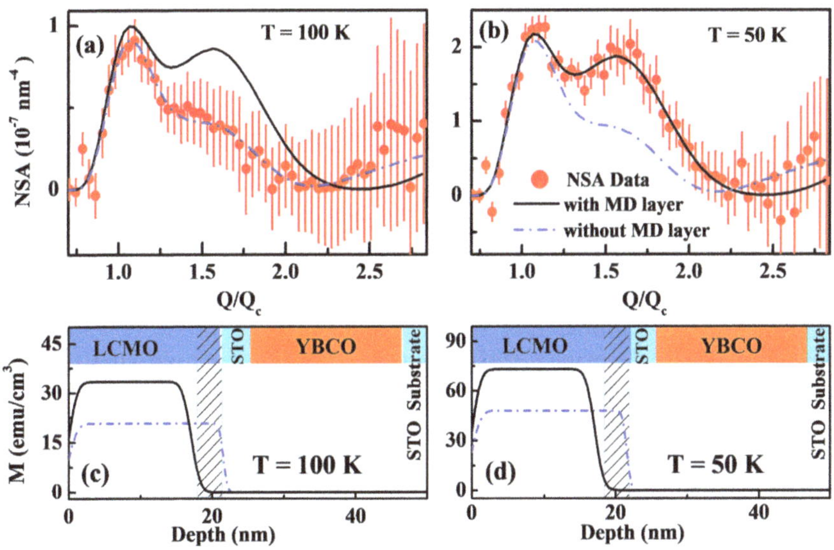

Figure 3.12. NSA data with fits assuming different magnetic depth profiles from a YBCO/STO/LCMO heterostructure at (a) 100 K and (b) 50 K. Different magnetization (M) depth profiles used to fit NSA (PNR) data at (c) 100 K and (d) 50 K, respectively. The results suggest the formation of a magnetic depleted layer at the STO/LCMO interface below the superconducting transition temperature. Reproduced with permission from [180]. Copyright 2016 The Royal Society of Chemistry.

superconducting order parameter into the FM layer, which modifies the magnetization at the interfacial region. Later Paull *et al* [182] also confirmed the magnetic modulation in the YBCO/STO/LCMO heterostructure below T_{SC} and they also showed that the magnetism of the depleted layer can be restored by applying a magnetic field that partially destroys the superconductivity in YBCO.

Recently, LSMO (FM)-based YBCO/LSMO and YBCO/STO/LSMO systems were also studied for understanding magnetic modulation below T_{SC} [183–185]. Systematic studies show formation of the MD layer both in direct proximity hybrid structures (YBCO/LSMO) as well as in the tunnelling geometry hybrid systems (YBCO/STO/LSMO) below T_{SC} (~65 K). PNR results below T_{SC} suggested the emergence of a marginally negative magnetization (magnetization opposite to the applied field) in the LSMO layer over a thickness of ~60 Å at the YBCO/LSMO interface in the YBCO/LSMO bilayer and of thickness 35 Å at the STO/LSMO interface in the YBCO/STO/LSMO trilayer heterostructures below T_{SC}, as a result of MPE. PNR results also showed that the thickness of the interfacial LSMO MD layer decreases upon increasing the thickness of the insulator (STO) layer, similar to the previous experiment [180]. This variation of the MD layer with the thickness of the insulating layer suggests that the tunnelling of Cooper pairs is a possible mechanism for magnetization modulation below T_{SC}. The length scale associated with this tunnelling-driven proximity effect strongly suggests the presence of long-range Cooper pairs in this system. Further, the dependence of the MPE on the superconducting state of the YBCO layer in the YBCO/LSMO (direct) and YBCO/STO/LSMO (tunnelling) hybrid structure was studied by reducing the T_{SC} through ion irradiation [184]. The evolution of magnetization at the interfacial FM layer, measured using PNR as a function of temperature for both the irradiated heterostructures showed the absence of the proximity effect observed in the unirradiated heterostructures. The absence of a proximity effect was attributed to the suppression of the superconductivity, as seen in macroscopic magnetization measurements (SQUID) of the heterostructures after ion irradiation. The stacking sequence is another key parameter for oxide heterostructure which can impact both the strength (magnetization) as well as the nature of the emergent phenomena at the interfaces. The stacking sequence, in addition to thickness, type of substrate, and annealing conditions in YBCO/LSMO heterostructures is also known to affect lattice strain as well as cause defect pinning and thus can influence the properties such as T_{SC}, magnetization, critical current density, and MPE, etc. Recently Bhatt *et al* [185] studied the effect of the stacking sequence on magnetic and superconducting properties in LSMO/YBCO and LSMO/STO/YBCO heterostructures, which consequently may affect the MPE. The stacking sequence dependence of MPE has been investigated using spin-dependent PNR experiments. An increase in the T_{SC} and magnetization was found for both of the heterostructures as compared to heterostructures with reversed stacking order studied earlier [183, 184]. Evolution of magnetization at the interface of the FM LSMO layer was obtained as a function of temperature for both heterostructures. A decrease in MPE-induced MD layer thickness for heterostructures with higher T_{SC} was observed. A comparison of different studies with the present results suggested that competition between the average magnetization of the FM and transition temperatures of the SC are important parameters that dictate the strength of the proximity effect due to the complex interaction of SC and FM in these systems [185].

A comparison of PNR results from YBCO/STO/LSMO hybrid system under three conditions: (i) trilayer heterostructure with stacking of YBCO/STO/LSMO grown on MgO substrate (nominal structure MgO (substrate)/YBCO/STO/LSMO) [183], (ii) ion irradiation of the heterostructure mentioned in (i) [184], and (iii) trilayer heterostructure with reverse stacking sequence (MgO/LSMO/STO/YBCO; trilayer grown on MgO substrate) [185], are shown in figure 3.13. The three heterostructures studied in the three cases mentioned above are named S1, S2, and S3, respectively, and the schematic of the heterostructures is shown in the inset (i) of figure 3.13(a). The inset (ii) of figure 3.13(a) shows the temperature-dependent normalized ZFC magnetization ($M(T)$) data for three multilayers, indicating widely different values of T_{SC} with a maximum T_{SC} of 85 K for S3 (reverse stacking) and the lowest T_{SC} of 30 K for the irradiated sample. Figure 3.13(a) shows the XRD data from three samples that suggest a lattice expansion (contraction) for the YBCO layer along the c-axis for ion-irradiated (reverse stacking) heterostructure, which is contributing to the change of T_{SC} in three systems. The inset (iii) of figure 3.13(a)

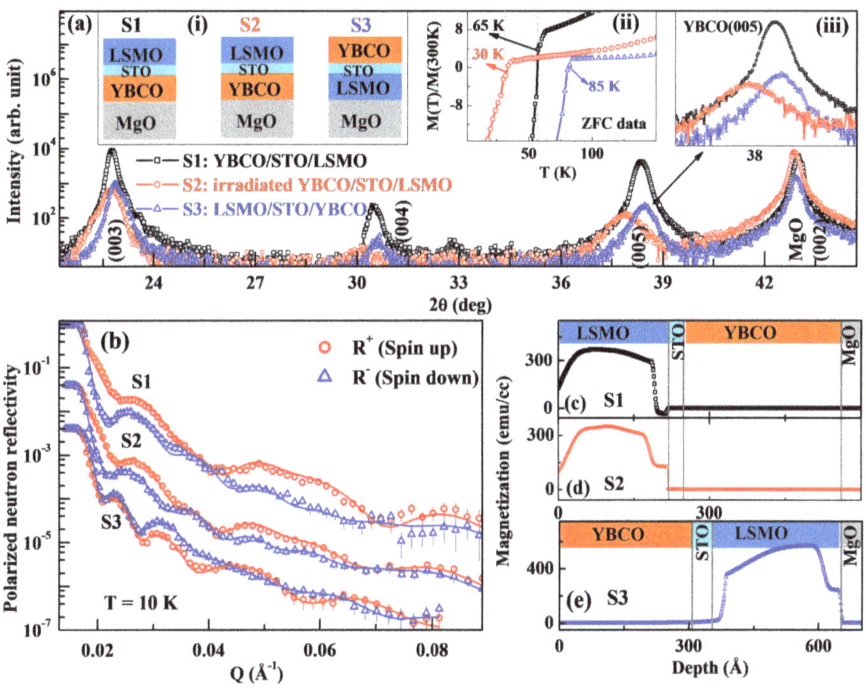

Figure 3.13. (a) XRD data from three YBCO/STO/LSMO heterostructures grown on MgO substrate and the schematic of the heterostructures are shown in the inset (i). Insets (ii) and (iii) of (a) show normalized ZFC $M(T)$ data and a magnified version of the (005) reflection of YBCO, respectively, for three heterostructures. (b) Spin-dependent PNR data and corresponding fits for the heterostructures at 10 K with magnetization depth profiles obtained from PNR data for the three heterostructures (c) S1, (d) S2), and (e) S3.

shows the magnified version of (005) YBCO Bragg reflection. XRR measurements were also used to obtain the detailed layer structure of the heterostructures. Spin-dependent PNR data (symbol) with the corresponding fits (solid lines), which are vertically shifted for better visualization, for three heterostructures, are shown in figure 3.13(b). The magnetization depth profiles obtained for these heterostructures from PNR data are shown in figures 3.13(c)–(e).

These results suggest a similar crystalline structure for all three samples with different macroscopic strains for the heterostructures grown with different stacking sequences, e.g. S1 [MgO/YBCO–STO–LSMO] and S3 [Mgo/LSMO–STO–YBCO]. However, magnetization depth profiles obtained from PNR below T_{SC} show a reduced (~more than 40%) length scale for MD layer formation in the interfacial LSMO layer for the LSMO–STO–YBCO heterostructure (tunnelling geometry) as compared to that of the YBCO–STO–LSMO (reverse stacking sequence) heterostructure. A higher reduction in the thickness MD layer was observed in the case of the bilayer heterostructure in direct proximity geometry. Irradiation of the sample (S2) caused reduction of its T_{SC} to 30 K, which wiped out MPE in this sample (figure 3.13(d)). Thus, different studies on FM/I/SC and FM/SC hybrid structures suggested that the average magnetization and transition temperatures of the FM and SC are important parameters that dictate the strength of the proximity effect due to the complex interaction of SC and FM in these systems.

3.7 Strain-driven interfacial magnetism in complex oxide heterostructures

Strain is a widespread phenomenon related to the synthesis, fabrication, and application of all types of materials. The effect of strain on the properties of materials has been studied for many centuries using pressure (hydrostatic strain) as a parameter. We are all aware of the role of macroscopic pressure on a body as one goes deeper into the Earth or the ocean. Pressure can dramatically change many physical properties and the functionality of materials and biological entities. Strain is one of the key parameters for understanding many physical phenomena at the nanoscale. Using epitaxy in thin oxide films, it is possible to achieve a strain far beyond the value where they would normally crack in bulk. Controlling the electronic and magnetic properties of complex oxide materials employing external stimuli, such as strain, is critical for the design of functional devices. In 2012 Hwang *et al* [186] advised that due to the strong correlation between charge, spin, and orbital degrees of freedom in complex oxides, strain can profoundly affect magnetism. Strain modifies the tilting of metal–oxide octahedra, thus influencing the occupation of orbitals and spin–orbit coupling. Such effects can manifest in emerging anisotropy, ordering temperature, or saturation magnetization of the material. Indeed, epitaxial strain is often thought to be the origin of uniaxial anisotropy in some oxides. Collective interactions among charge, spin, orbital, and lattice order can lead to complex behaviour in doped manganites, such as colossal magnetoresistance (CMR) and metal–insulator transitions (MITs), that may result from the coexistence of ferromagnetic, metallic, and insulating phases.

These interactions can be greatly influenced by the environment, e.g. magnetic field, light, stress, and disorder, and thus can be tuned for broad applications.

The role of strain in manganite thin films has been identified as an interesting perturbation for a basic understanding of physical interactions and has been proposed as an adequate mechanism to describe the multiscale multiphase coexistence observed in manganites [187]. However, unlike bulk manganites, in which compressive stress mostly favours the ferromagnetic phase, theoretical and experimental studies of manganite films paint a decidedly mixed picture and the role of strain and magnetism in manganite films is far from understood. The epitaxial strain in complex oxide heterostructures has been varied either by using substrates with a lattice mismatch or by varying the film thickness. However, the study of systematic intrinsic strain effects using these methods is difficult because it is not possible to discriminate the effect of many other variables such as lattice relaxation, stoichiometry, crystalline quality, and microstructure. Figure 3.14(a) shows the temperature dependence of normalized magnetization of LSMO films epitaxially grown on different substrates, suggesting different saturation magnetization as well as different magnetic transition temperatures (Curie temperature, T_c) [188]. Figure 3.14(b) shows the variation of T_c with epitaxial strain for LSMO film, suggesting a mixed report, i.e. both compressive and tensile epitaxial strain can decrease the T_c.

To overcome the difficulty of tailoring epitaxial strain, another approach (piezoelectric strain) was used where it is possible to vary the strain state of thin films *in situ* while the negative variables could be kept fixed. In the literature, the piezoelectric strain on epitaxial film has been applied by two methods. In the first method the structural phase transitions of ferroelectric (FE) (in particular $BaTiO_3$, BTO) substrates are used to generate intrinsic strain in manganite films. A large biaxial strain can be produced at the transition temperatures, which modify the electronic and magnetic properties of manganite thin films purely due to strain. In the case of another method, the manganite films are deposited on a piezoelectric single-crystal substrate and strain is induced via the converse piezoelectric effect (strain is generated due to an applied electric field in the piezoelectric substrate).

Another approach involves using a mechanical jig to apply pure bending stress to a film [189]. Unlike epitaxial and piezoelectric strain cases, which require epitaxy between the film and substrate, a mechanical jig can apply stress to any film/substrate combination, only requiring that the substrate should be reasonably thin (<500 μm thick) to allow bending. The inset of figure 3.15(a) shows a schematic of applying uniform bending stress, using a four-point bending jig, across a large area sample (e.g. 1 cm^2 sample) suitable for neutron scattering. Even though the substrate is 'thin', its thickness is still orders of magnitude greater than that of the film. Therefore, the resulting strain in the film is uniform as a function of depth. Singh *et al* [189] measured simultaneously the transport and PNR data for $La_{0.27}Pr_{0.40}Ca_{0.33}MnO_3$ (LPCMO) grown on a (110) NGO substrate as a function of temperature and bending stress. PNR measurements provided the magnetization depth profiles at saturation across an LPCMO film as a function of the exclusive application of elastic stress and suggested strong coupling between strain and magnetization in the film. This study is a case of a single sample being stressed

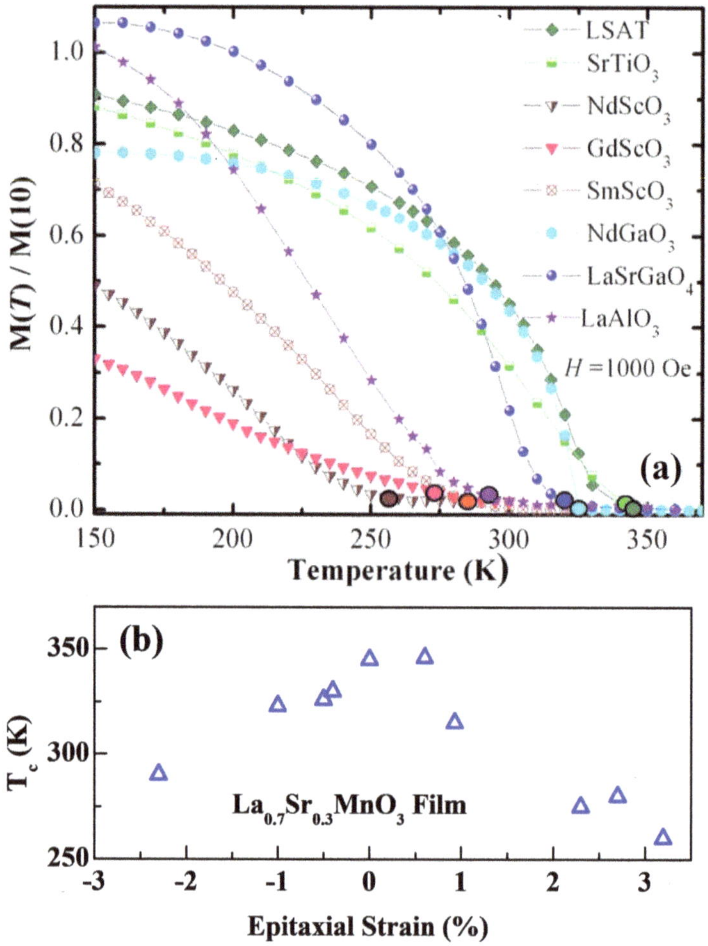

Figure 3.14. (a) Temperature dependence of magnetization, normalized at 10 K, for LSMO film on different substrates cooled in a 1000 Oe field. Closed circles indicate the Curie temperature, T_C. Reproduced with permission from [188]. Copyright 2009 AIP Publishing. (b) Magnetic transition temperature (T_C) of LSMO manganite thin films vs in-plane epitaxial strain grown on different substrates. The results suggest a mixed report on strain-dependent magnetic properties.

well within the elastic regime, and it unequivocally determined the exclusive role of stress on the saturation magnetization in a thin film.

Figure 3.15(a) shows the transport ($R(T)$) data for the LPCMO film under an applied strain (ε) in the range of 0% (no strain), −0.011% (compressive), and +0.011% (tensile) along the in-plane ($1\bar{1}0$) direction of NGO (magnetic easy axis) and in a magnetic field of 6 kOe (applied along the easy axis). PNR measurements as a function of applied stress were carried out at a constant temperature of 78 K while cooling across the metal to the insulator transition temperature. Figure 3.15(b) shows the PNR data for LPCMO film at 78 K (while cooling) under three values of the applied strains and the corresponding magnetization (M) depth profiles obtained

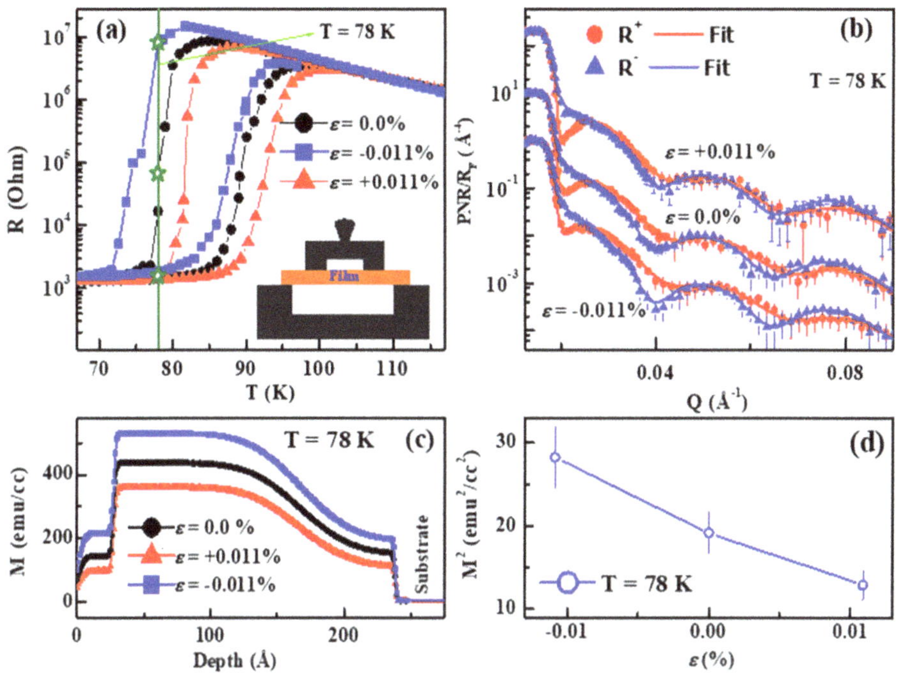

Figure 3.15. (a) Transport data for $La_{0.27}Pr_{0.40}Ca_{0.33}MnO_3$ (LPCMO) grown on (110) NGO substrate on the application of applied bending strain (ε) of 0% (without strain), +0.011% (tensile strain), and −0.011% (compressive strain). The inset shows a schematic of bending strain. (b) PNR data for the LPCMO film on the application of applied different bending strains at 78 K, during the cooling cycle. (c) Magnetization depth profiles for the LPCMO film at T = 78 K for different cases of strain as obtained from PNR data. M^2 vs ε curve for LPCMO film. Reproduced with permission from [189]. Copyright 2012 the American Physical Society.

from PNR data are shown in figure 3.15(c). The PNR analysis showed that compressive strain increases the average magnetization while tensile strain suppresses the average magnetization. The M^2 versus ε curve at 78 K is shown in figure 3.15(d), which was used to calculate the coupling coefficient for strain and magnetism in this system [189].

The influence of applied elastic bending strain on the magnetization depth profile of a $La_{0.8}Sr_{0.2}MnO_3$ (LSMO) epitaxial film (thickness ~50 nm) grown on (001) STO was also studied using PNR, where PNR measurements were performed as a function of applied elastic bending strain (ε) of ~±0.03% [190]. The magnetization depth profiles as a function of applied strain suggested a small change in the magnetization of the LSMO film. Thus a negligible coupling of elastic bending strain with magnetization was observed for the LSMO film as compared to the LPCMO film, which was attributed to weak and negligible phase coexistence in the LSMO film. The mechanical bending jig is capable of providing a tunable and reversible bending uniaxial stress which can easily be deployed in PNR instruments for measuring the magnetization of heterostructures. The bending approach further eliminates the defects and other external factors as possible origins for phase coexistence in manganite thin films. Unlike other studies for controlling the phase–property correlation via changing the film thickness, substrate orientation, interfacial coupling, or chemical doping, this tuneable bending stress provides an effective way to dynamically modify phase competition and understand the complex phase diagram in a single manganite film.

3.8 Emergent and interface-induced magnetism at complex oxide interfaces

Perovskite (ABO_3) complex oxide heterostructures with dissimilar *B*-site cations make *B–O–B′* bonds at interfaces and exchange interaction across this can deviate from that found within the adjoined materials with *B–O–B* and *B′–O–B′* bonds. Depending on this interaction at the interface, the *B–O–B′* coupling can be ferromagnetic even when both *B–O–B* and *B′–O–B′* interactions are antiferromagnetic. Ueda *et al* [191] tuned the *B–O–B*' interaction in $ABO_3/AB'O_3$ (where ABO_3 and $AB'\ O_3$ were $LaCrO_3$ and $LaFeO_3$, respectively) superlattices by growing these superlattices along different crystallographic orientations for studying long-range FM or AFM phenomena at interfaces. The existence of different magnetic orders at the complex oxide interfaces different from their constituent oxides have been reported, e.g. ferromagnetism at interfaces between two different AFM oxides [192–194], and between a paramagnetic metal and AFM insulator [161, 195–198]. The PNR technique, due to its intrinsic strength in deciphering interface magnetism, has played an important role in revealing the interfacial magnetic phases as well as its coupling with different oxide layers.

Macroscopic magnetometry (SQUID) and resonant x-ray magnetic scattering studies discovered ferromagnetism in CMO/CRO superlattices [198], where CMO is an AFM insulator and CRO is a paramagnetic metal. Theoretically, the ferromagnetism in the CMO/CRO heterostructures was explained in terms of a ferromagnetic double exchange interaction among Mn ions in one unit cell of CMO due to canted AFM [196]. A finite magnetization with the EB effect for a CMO/CRO superlattice was investigated using the depth-sensitivity of the PNR and it was demonstrated that the interfacial FM is confined to a single unit cell of CMO [161]. The PNR results also confirmed the existence of ferromagnetism in one unit cell of CMO at both the CMO/CRO (CMO on CRO) and CRO/CMO (CRO on CMO) interfaces. In another important study using PNR the existence of an FM phase at the interface between a paramagnetic metal of $LaNiO_3$ (LNO) and CMO was revealed [195]. PNR results clearly indicated the existence of FM in one unit cell of CMO at both the CMO/LNO (CMO on LNO) and LNO/CMO (LNO on CMO) interfaces, which is dependent on the metallic nature of LNO [195]. Theoretical analysis suggested that the strong dependence of the FM on the conducting state of LNO in LNO/CMO superlattice was indicative of an interfacial double exchange interaction mediated by the LNO e_g band [195].

Several fascinating interfacial phenomena that are driven by spin, charge and orbital reconstructions at the interface have also been observed in heterostructures involving oxide FMs such as LSMO and oxide AFMs such as BFO. Theoretical calculations have suggested a weak FM for BFO (~0.03 μ_B) as a consequence of canting of the AFM structure due to DM interaction. Interestingly, an enhanced magnetic moment in the BFO layer at the interface was observed, which was attributed to Fe–Mn hybridization and orbital reconstruction. Recent PNR measurements [159, 199, 200] from BFO/LSMO heterostructure have suggested a magnetization of ~100 emu/cc for the interfacial BFO layer of thickness ~25–35 Å. However, the PNR study also suggested the existence of different magnetic couplings: FM and AFM in this system [159]. Different coupling at the interfaces might have resulted due to different surface terminations. *Ab initio* calculations also confirmed both AFM and FM exchange coupling across the BFO/LSMO interface, suggesting surface termination dependence of magnetism in LSMO film [199].

Using PNR measurements, Bhatt *et al* [160] found the emergence of the interfacial ferromagnetic LNO layer of thickness ≈ 22–35 Å (≈ 6–9 unit cell) and magnetization ≈ 80 emu/cc at the LNO/LSMO interface for $T < 100$ K. The PNR results also show both FM and AFM coupling between the LSMO and interfacial LNO layers, which may be dependent on the stacking order. The results also suggested that the stacking order of the layers plays a significant role in deciding the magnitude and sign of the interfacial exchange interaction in these heterostructures. Therefore, for the investigation of emerging interfacial magnetism in these oxide heterostructures, the PNR technique, due to its inherent strength to reveal small magnetism at interfaces with sub-nanometre depth resolution, has been and will continue to be playing a unique and major role in this exciting research area.

3.9 Superdense and nonmagnetic Co phase at interfaces

3*d* transition metals, e.g. Fe, Co, and Ni show, interesting magnetic properties both in their bulk and thin-film form. Co, which shows a hexagonal closed pack (*hcp*) structure, is also well-studied theoretically. Theoretically, the magnetic moment of Co decreases with decreasing atomic volume and the magnetic moment of a face-centred-cubic (*fcc*) Co phase nearly vanishes when the atomic volume is reduced by a factor of ~1.4. A high-pressure experiment on bulk Co has shown that the *hcp* to *fcc* phase transition occurred around this atomic volume, corresponding to a density of ~1.4 times the normal Co density. However, there was no experimental study in the literature for Co, showing highly dense nonmagnetic properties. Recently Banu *et al* [201, 202] studied the depth-dependent structure and magnetic properties of a Co thin film. The polycrystalline Co film was deposited on a piranha solution cleaned and HF etched Si [111] substrate using an electron-beam deposition technique. The depth-dependent structure and magnetism at RT were obtained using XRR and PNR. Figure 3.16(a) shows XRR data (symbols) along with fit (solid line) from the Co film. The ESLD depth profile (figure 3.16(b)) of the film obtained from XRR data suggested the presence of high-density Co layers at the film–air and film–substrate interfaces. This was also confirmed from NSLD data obtained from PNR (figure 3.16(d)). PNR data (symbols) along with fits (solid lines) from the Co film at room temperature (RT) are shown in figure 3.16(c). Figure 3.16(d) shows the NSLD and MSLD depth profile of the film, which best fitted the PNR data at RT. These

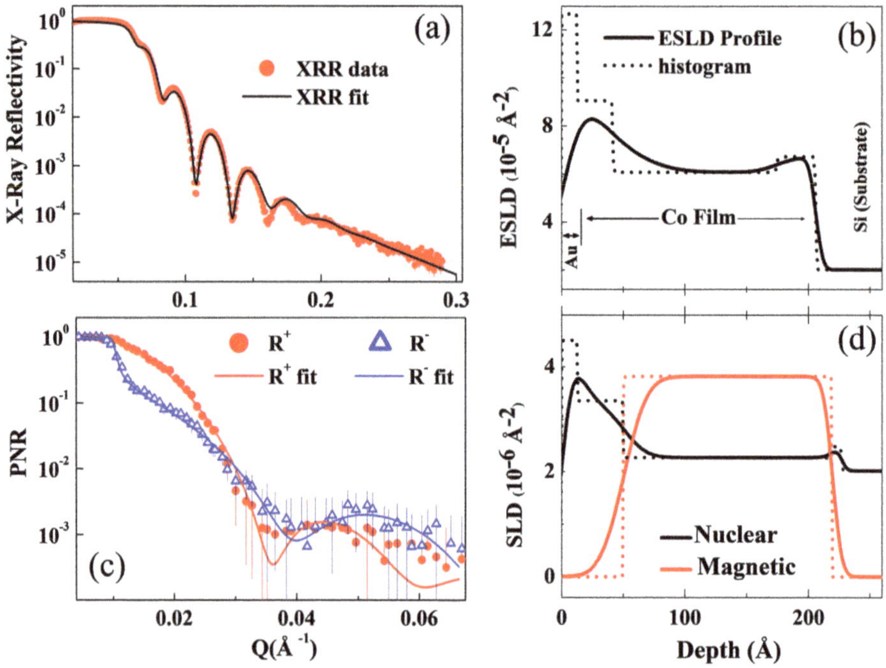

Figure 3.16. (a) X-ray reflectivity measurements from Si (substrate)/Co (23 nm)/Au (2 nm) heterostructure. (b) Electron scattering length density (ESLD) depth profile of the heterostructure obtained from XRR. (c) Polarized neutron reflectivity measurements at room temperature from the heterostructure. (d) Nuclear and magnetic SLD depth profile of the heterostructure obtained from PNR. The dashed lines in (b) and (d) show the histogram of SLD profiles indicating different layers and interfaces. Reproduced with permission from [202]. Copyright 2018 IOP Publishing.

high-density layers were about ~30 Å thick at the interfaces. Most interestingly, PNR data also suggested that the high-density layers at the interfaces were nonmagnetic, which can be seen in the MSLD (figure 3.16(d)) depth profile obtained from PNR. It was also confirmed from high-resolution cross-sectional transmission electron microscopy, suggesting that these high-density layers at interfaces showed an *fcc* structure unlike bulk Co, which has an *hcp* structure. Thus, both the XRR and PNR experiments confirmed that there was a high-density Co layer present on both interfaces (film–air and substrate–film) of the film. As mentioned earlier such a high-density Co phase had only been observed under high pressure. It was conjectured that the migration of Co atoms along grain boundaries during deposition might cause such high pressure and give rise to such a phase. Several Co films with different thicknesses grown on different substrates were investigated using XRR and PNR techniques and the results suggested that the growth of high-density nonmagnetic cobalt phases in the thin film depends on (i) a capping layer on the cobalt film, (ii) the thickness of the cobalt film, and (iii) the nature of the substrate on which the cobalt film is grown [202]. This unique result on the formation of superdense nonmagnetic Co *fcc* phase at interfaces of the film vindicated the strength of PNR in detecting such ultra-thin layers and their magnetism.

3.10 Tracking interdiffusion and self-diffusion kinetics at interfaces

Heterostructure systems provide model systems for researchers to investigate interface properties with sub-nanometre resolution, e.g. stabilization of non-equilibrium structures of interface layers and their strain and morphology. Interactions related to electronic and magnetic coupling between layers give rise to novel magnetism and transport behaviour. In addition, annealing of metal–metal heterostructures provides an excellent route to produce intermetallic alloy phases at interfaces and it becomes very important in the case of ultra-thin films where alloying occurs at much lower temperatures compared to the melting temperatures of its components in bulk. Therefore, studies related to interdiffusion in compositionally modulated heterostructures are of interest for the formation of alloys at low temperatures and observation of solid-state reaction (SSR) at interfaces in thin films has also added to the importance of investigations related to interdiffusion in these materials. The diffusivity associated with the interdiffusion of atoms at interfaces in thin-film heterostructures is very low (10^{-19}–10^{-23} cm^2 s^{-1}) and is associated with a diffusion length on the sub-nanometre length scale. XRR and PNR are two complementary techniques to investigate such very low diffusivity of atoms at interfaces in thin films [13–15, 17, 203]. In addition, the formation of an alloy also modifies the interactions at interfaces and consequently modifies the structure and magnetic properties of the heterostructures. The process of alloy formation may further influence the diffusivity of atoms across the interface, either aiding or deterring further alloy formation. The diffusion of the elements at interfaces and the rate of alloying at interfaces depends on several physical parameters: the interface roughness, growth sequence, crystallinity of the layers, and surface energy of the components. Annealing may also lead to redistribution of defects and rearrangement of bond lengths and bond angles, as well as reduction of strain, which results in a drastic change in many microscopic physical properties, e.g. the vibrational, electronic, magnetic, and mechanical properties as well as the thermodynamics of phase formation. Self-diffusion is a basic process in a single component system, which controls material properties and is a very important component of diffusion studies in thin films. Depth profiling techniques such as radioactive tracer and secondary ion mass spectroscopy (SIMS) were commonly used for self-diffusion measurements in thin films. In the last decade reflectometry techniques (XRR and PNR) have been successfully used to study self-diffusion in many systems due to their large sensitivity to thin-film interfaces [204].

XRR and PNR techniques have been used in the literature extensively for precise measurement of diffusion/interdiffusion at the interfaces from compositionally modulated multilayers [13–15, 203, 205]. In these techniques the intensity of the Bragg peaks of a multilayer is monitored as a function of annealing temperature and time. Diffusion takes place across the interfaces of a multilayer upon annealing at an elevated temperature. The interfacial modification due to interdiffusion reduces the intensity of the Bragg peaks in the reflectivity pattern, due to the reduction in contrast at interfaces. Diffusivity at the interfaces in a multilayer usually results in a shorter diffusion length of ~tens of Å and reflectometry techniques are very useful to study the diffusion at the microscopic level in thin films and multilayers [17, 203].

Reduction in the intensity of the Bragg peak on annealing provides an estimate of the diffusivity and hence activation energy of the constituent (Fe, Pt, or Cu) in multilayer. Diffusivity (D) can be estimated using the expression [17, 203]

$$\ln I(t)/I(0) = \frac{-8\pi^2 n^2 Dt}{d_{\mathrm{bl}}^2}, \tag{3.10}$$

where $I(t)$ and $I(0)$ are the intensities (reflectivities) of the nth order Bragg peak after annealing for time t, and d_{bl} is the bilayer thickness in the as-deposited state. Diffusion length averaged over three dimensions, describing the average displacement of atoms during annealing in a time interval t, is given as, $l_d = \sqrt{6Dt}$, where D is the diffusion constant. Thus a reduction in Bragg reflectivity is used to calculate the diffusivity in the continuum approach via equation (3.10) and assuming Arrhenius behaviour, the temperature dependence of the diffusivity provides activation energy (E_a), given by $D = D_0 \exp(-E_a/k_B T)$. ln(D) versus $1/T$ plots are in general a linear function in the entire temperature interval and the linearity implies that there is only one activation energy associated with the process.

Figure 3.17(a) shows the XRR data for a limited Q-range for a FePt/Cu multilayer in pristine and annealed (at different temperatures) conditions [206]. Reduction of reflectivity at the Bragg peak (denoted by vertical arrow) as a function

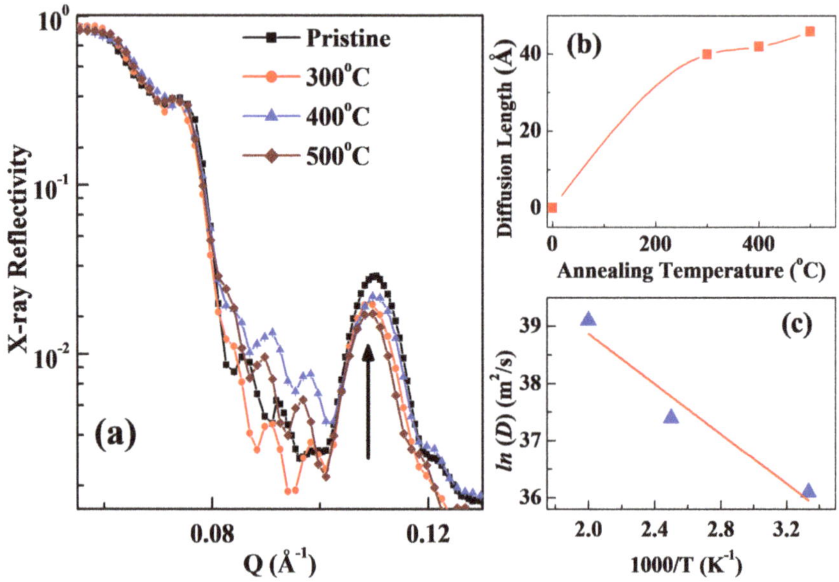

Figure 3.17. (a) X-ray reflectivity data from an FePt/Cu multilayer sample for different annealing temperatures. The multilayer was annealed at different temperatures under a vacuum for a fixed time of 1 h. The reduction in reflectivity of the Bragg peak (indicated by a vertical arrow) at different annealing temperatures was used to estimate the diffusion length at interfaces. (b) Diffusion length as a function of temperature obtained from XRR data. (c) Arrhenius plot of diffusivity for an FePt/Cu multilayer. The activation energy obtained from this plot was 0.018 eV, suggesting faster interdiffusion at the interfaces. Reproduced with permission from [206]. Copyright 2020 AIP Publishing.

of annealing temperature is evident from figure 3.17(a), which was used to estimate the diffusivity and diffusion length of atoms at interfaces. Figure 3.17(b) shows the variation of diffusion length with annealing temperature. A diffusion length of ~40 Å was obtained on annealing the multilayer at 300 °C, suggesting higher diffusion at interfaces in this stage but further annealing of the multilayer at 400 °C and 500 °C, a marginal increase in diffusion length (~44 Å) was found. A significantly lower diffusion of atoms at interfaces on annealing at higher temperatures of 400 °C and 500 °C suggests a kind of blocking effect, which inhibits further atomic diffusion at interfaces. Diffusivity at different annealing temperatures is plotted as an Arrhenius plot for FePt/Cu multilayer in figure 3.17(c) and activation energy (E_a) was estimated to be about 18 meV. The reflectivity techniques are very effective to estimate the diffusion length and activation energy of atoms at interfaces. Interfacial diffusion are very important for controlling interface structure and morphology in the heterostructures for technological application. Depth-sensitivity and resolution of these complementary techniques (XRR and PNR) become even more useful in the investigation of interface structure and morphology, particularly in many oxide heterostructures where direct methods (such as TEM with EELS) cannot provide information with sufficient resolution.

It is noted that the XRR cannot be used to study the self-diffusion of a single atomic species because x-ray scattering cannot differentiate any mixing of atoms of the same species. However, the neutron has different nuclear scattering lengths for isotopes of the same element and it can thus distinguish isotopes and can be used to study self-diffusion using isotopically distinct layers of the same atomic species. PNR has been used extensively for self-diffusion studies in the crystalline as well as amorphous phases of isotopically graded multilayers. Reflectometry techniques allow the tracking of low diffusivity due to their excellent depth resolution. Here we will discuss the formation of intermetallic alloy on annealing of several technologically important multilayer systems, namely Ni/Ti, Ni/Al, and FePt/Cu.

Another important and related issue in metallic systems is the identification of the intermetallic alloy phase formed on the annealing of metallic multilayers, which can drastically change their physical and magnetic properties. In the case of thin-film multilayers, the diffusion/interdiffusion mechanism couples with layer thicknesses and the reactive phase formation follows a sequential reaction pathway that allows only one phase to form at a given annealing temperature and annealing time. Thus, it becomes essential to estimate the stoichiometry of the phase evolved at interfaces on annealing and understand the kinetics. In general, x-ray diffraction (XRD) and grazing incidence XRD (GIXRD) are used to see the evolution of crystalline phases on annealing but due to experimental constraints (sensitivity for ultra-thin interface alloy) it is often difficult to predict the actual composition of the alloy formed at interfaces in the sub-nanometre length scale using these techniques. XRR and NR have been successively used together as two independent measurements to estimate the actual composition of an alloy phase evolved as a function of annealing in multilayers [13–15, 17, 205]. A comparison of ESLD and NSLD values, obtained from XRR and PNR measurements, respectively, for the alloy phase, provides an estimation of the exact composition of the alloy phase [13–15, 17, 205].

3.10.1 Ni/Ti multilayer

Ni/Ti-based systems have attracted much attention in engineering design as a promising pair to develop a novel combination of physical and mechanical properties as well in many emerging areas, including neutron mirrors and soft x-ray optics. Ni, among the metals, has a large positive scattering length and Ti has a negative scattering length for thermal neutrons. The large difference between the nuclear SLD of Ti and Ni makes the Ni/Ti multilayer an ideal pair with large contrast for the fabrication of neutron supermirrors as a neutron optical component. The Ni/Ti multilayer is also considered a model system for demonstrating the solid-state amorphization (SSA) reaction, where alloying at interfaces due to annealing progresses through amorphization. The formation of an amorphous phase on annealing arises due to large asymmetry in atomic mobility of the constituents of the system and large negative Gibbs free energy of mixing. Contrary to alloying due to the presence of an amorphous phase at interfaces, Singh *et al* [13] found alloying at the interfaces of the Ni/Ti multilayer on annealing, which progresses through the formation and growth of microcrystals. Ni–Ti alloys have become technologically important materials due to their excellent combination of many attractive properties such as the shape memory effect (SME), superelasticity (SE), high damping capacity, biocompatibility, corrosion resistance, and excellent mechanical properties.

The formation of an alloy phase at the interface and its influence on the electronic properties of the Ni/Ti multilayer has been studied using core-level photoemission, XRR, and XRD techniques. Combining XRD and XRR in specular and off-specular mode and PNR measurements, Singh *et al* [13] studied the evolution of the alloy phase at interfaces on annealing Ni/Ti multilayers at a temperature of 300 °C and 400 °C, for structure, magnetic, and morphological properties. PNR data for the pristine (as-deposited) Ni/Ti multilayer and a multilayer annealed at 400 °C is shown in figure 3.18(a) (left panel). A larger separation between spin-dependent neutron reflectivity for the pristine multilayer as compared to the annealed multilayer suggests higher magnetization for the Ni layer in an as-deposited system. The depth-dependent NSLD and MSLD profiles obtained from the PNR data are shown in the middle panel of figure 3.18(a). The as-deposited sample indicates a periodic structure, evident in the NSLD and MSLD depth profiles. Post-annealed NSLD and MSLD clearly indicate the shrinking of the Ni and Ti layers due to alloying at the interfaces. XRR and PNR data revealed the formation of the Ni_3Ti alloy phase at the interfaces due to interdiffusion of atoms [13]. Tracking the reflectivity at the Bragg peak provided the diffusivity and diffusion length at different temperatures of annealing. The variation of growth of the alloy layer thickness, diffusion length, and reduction of Ni and Ti layer thickness is also shown in the bottom panel of figure 3.18(a). The reduction in the thickness of the Ni layer and Ti layer and the corresponding growth of the alloy layer thickness also resulted in the reduction of the magnetic moment of the remaining Ni layer. PNR results also suggested that the alloy layer is nonmagnetic [13]. Interestingly, this study also suggested that alloying at the interfaces had progressed through the formation of alloy crystallites unlike

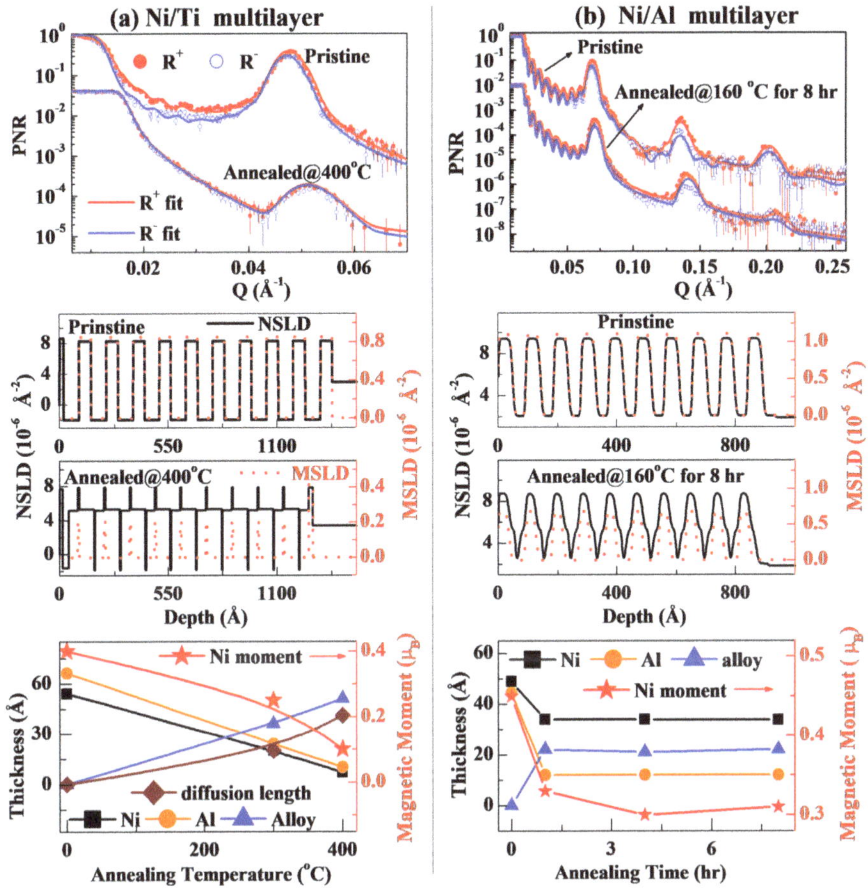

Figure 3.18. (a) PNR data (symbols) and corresponding fits (solid lines) for an Ni/Ti multilayer in pristine and annealed (@ 400 °C) conditions. For a better visibility, data have been shifted vertically for these two conditions. The middle panel below the PNR data shows the NSLD and MSLD depth profiles of the Ni/Ti multilayer for pristine and annealed conditions. The bottom panel shows the variation of other parameters as a function of the annealing temperature obtained from PNR measurements. (b) PNR data (symbols) and corresponding fits (solid lines) for a Ni/Al multilayer in pristine and annealed (@160 °C for 8 h) conditions. For a better visibility, data have been shifted vertically for these two conditions. The middle panel shows the NSLD and MSLD depth profiles of the Ni/Al multilayer for pristine and annealed conditions. The bottom panel show the variation of other parameters as a function of the annealing temperature obtained from PNR measurements.

several other earlier studies on Ni/Ti multilayers and such a rapid alloying is possible through grain boundaries and defects in the nanocrystalline alloy phase.

3.10.2 Ni/Al multilayer

Ni and Al both have fcc structures at room temperature with cell parameters 3.52 Å and 4.05 Å, respectively, with Al having a larger atomic radius. Al is soluble up to 20% in Ni and most importantly, apart from disordered solid solutions, Ni and Al form several ordered intermetallics. Ni/Al-based alloys and heterostructures are important systems for industrial application because of their high strength, hardness, high melting point, and resistance to oxidation along with favourable electronic and magnetic properties. Nickel aluminides form an important class of intermetallics for applications ranging from aeronautical industries to corrosion-resistant coatings. There are several methods such as cold rolling, ball milling or mechanical alloying, shock compaction, and self-sustained reactions for producing intermetallic alloys. However, thermal annealing of heterostructures in a vacuum is one of the suitable methods for inducing solid-state reactions between Ni and Al with short bilayer periods (~nm) to produce nickel aluminides. Reactions in the Ni–Al thin-film systems have been characterized extensively andwere studied before 1990 for the formation of aluminides, which has been summarized in a review by Colgan [207]. Various combinations of Ni and Al layers in contact with different aluminides, e.g. NiAl, $NiAl_3$, and Ni_3Al, were considered by Colgan [207] to study the formation of the phase at the interfaces during annealing and demonstrated that the first phase formed in these compounds is always $NiAl_3$. Using Auger electron spectroscopy (AES) for depth profiling and XRD, Rothhaar *et al* [208] studied different alloy phase formations after the annealing of an Al/Ni multilayer. Different alloy phases, Al_3Ni, Al_3Ni_2, and AlNi, formed depending on the annealing temperature (between 160 °C and 330 °C for 45 min).

The growth kinetics of the intermetallic alloy phase at the interfaces of a Ni/Al multilayer as a function of annealing time was systematically studied using x-ray scattering (XRD and XRR) and PNR [14]. The Ni/Al multilayer was annealed at a fixed temperature of 160 °C for different times (1–8 h) and the alloy formed showed remarkable stability with respect to annealing time, following an initial alloy formation at the interface. The stability of such multilayers is important for their applicability as corrosion-resistant coatings as well as metallization layers in microelectronic devices. In addition, XRR and PNR results for ESLD and NSLD, respectively, were used in two algebraic equations [14] for unique identification of the interface layer as the Al_3Ni intermetallic phase, which was magnetically dead. Figure 3.18(b) shows the spin-dependent PNR measurements (upper panel) for a Ni/Al multilayer in pristine and annealed at 160 °C for 8 h conditions. The depth-dependent NSLD and MSLD profiles obtained from the PNR data are shown in the middle panel of figure 3.18(b). XRR and PNR data revealed rapid (within the first hour of annealing) alloy ($NiAl_3$ phase) formation at the interfaces due to interdiffusion of atoms at the interfaces, which further blocks the interdiffusion and no increase in alloy thickness was observed on further annealing the multilayer for a longer time (~8 h). The bottom panel of figure 3.18(b) shows the variation of other parameters such as Ni moment, Ni, Al, and alloy layer thickness, with annealing time. This plot bears the signature of the 'blocking effect' as all the parameters mentioned above reach a saturation value (plateau) after the first hour of annealing. The PNR results also showed that the interface alloy layer was magnetically dead and the average magnetic moment in the Ni layer reduced from 0.52 to 0.30 μ_B per Ni atom due to Al diffusion into the Ni layers. Studies on Ni/Al heterostructures also suggested that the interface morphology influences the kinetics of the formation of alloy phases at interfaces. To see the effect of interface morphology on first-phase formation in the Ni/Al system, Swain *et al* [16] performed XRR and PNR measurements on a Ni/Al multilayer as a function of time at a fixed temperature of annealing and found different interface roughness/morphology at two interfaces, Ni/Al (Al layer on Ni layer) and Al/Ni (Ni layer on Al layer) for the pristine condition and the different roughnesses decided the first phase of alloy formation on annealing.

The reflectivity results confirmed rapid alloy formation at interfaces during an initial 1 h of annealing with different compositions of alloys, i.e. Al_3Ni and Al_3Ni_2, at the Ni/Al and Al/Ni interfaces, respectively; these interfaces showed different morphology in pristine case. The heat of formation and local densities of constituent elements are two other factors that also influence the formation of the first phase of the alloy upon annealing of multilayers [17]. Swain *et al* [15] studied Ni/Al multilayer systems with different thicknesses to see the effect of local densities on first-phase formation. Two multilayers comprising periodic bilayers of Ni and Al on Si substrate were grown by choosing the thickness ratio of Ni and Al such that the overall atomic ratios of Ni:Al were 3:1 and 1:3, which should form the intermetallics Ni_3Al and $NiAl_3$ if we make a homogeneous alloy from these multilayers [15]. The evolution of the alloy phase and the depth-dependent structure and magnetic properties as a function of annealing temperature was studied using the XRR, PNR, and SIMS techniques. The results showed that the first alloy phase formed on annealing of both the multilayers was $NiAl_3$. It was observed that the diffusion length decides the availability of the elements at the alloying front and changes the local density, favouring the phase which is kinetically favoured, i.e. $NiAl_3$ for both multilayers compared to the thermodynamically favoured phase (Ni_3Al) for one of the multilayers. This suggested that the definition of local density should be the material available within the diffusion length of the element under consideration. Thus, in the initial stage of diffusion, the growth of the interface alloy layer is decided primarily by the diffusion length, which is also responsible for deciding the stoichiometry of the alloy phase at interfaces. PNR results also suggested that the alloy phase is nonmagnetic.

3.10.3 FePt/Cu multilayer

The equiatomic ordered $L1_0$ or face-centred tetragonal (FCT) $Fe_{50}Pt_{50}$ (FePt) films have attracted considerable attention due to their properties such as high magneto-crystalline anisotropy, saturation magnetization, and high perpendicular magnetic anisotropy (PMA), which have potential applications in high-density recording media. However, a relatively high-temperature (>600 °C) annealing is necessary to transform the as-deposited disordered face-centred-cubic (fcc) phase (A1 phase) of FePt films into the ordered tetragonal $L1_0$ phase. The addition of Cu in FePt films is suggested as one of the effective ways to reduce the kinetic ordering temperature, which has been explored by forming $(FePt)_{100-x}Cu_x$ (FePtCu) ternary alloys. However, the addition of Cu in FePt modifies the magnetic properties of the system. Basha *et al* [205] performed a systematic study to investigate the evolution of structural and magnetic phases at interfaces of FePt/Cu multilayers on annealing at different temperatures for a fixed time (isochronal) and at a fixed temperature (600 °C) for different times (isothermal). XRR and PNR were used as the primary techniques to investigate the depth-dependent structure and magnetic properties, and diffusion-induced nucleation of the alloy phase at interfaces with isochronal and isothermal annealing (in vacuum) of FePt/Cu multilayers was also revealed. The isochronal annealing (in the temperature range of 300 °C–600 °C for a fixed time of 0.5 h) of the multilayers suggested a small interdiffusion of elements at the FePt/Cu (Cu on FePt) and Cu/FePt interfaces, which resulted in the suppression of magnetic properties. Upon isothermal annealing (at 600 °C) for longer times (1.5–6.5 h), large interdiffusion at the interfaces and formation of ternary alloy and iron silicide (at the substrate–film interface) phases were found. The results indicated that high-temperature annealing of the multilayers for a longer period produces an iron silicide phase at the substrate–film interface that has assisted in the evolution of the FePtCu ternary phase. The increased Cu content in the FePtCu ternary alloy film leads to phase coexistence with a larger fraction of the FCT phase, which is responsible for its low saturation magnetization with a high coercivity of 6 kOe at room temperature. These results demonstrate a mechanism for producing a ternary alloy phase of the designed composition by annealing the multilayers of different thicknesses, which showed distinctly different magnetic properties as indicated by PNR. The tuning of magnetic properties of ternary alloys by adding different contents of Cu may be desirable for future magnetic devices.

3.10.4 Self-diffusion in isotopic multilayers

In addition to studies related to interdiffusion of two different species in a binary multilayer (e.g. Ni/Ti, Ni/Al etc), self-diffusion in the nanometre length scales is also extremely important to tailor material properties as it controls various related kinetic processes and thus has been studied in multilayers of single chemical species but with different isotopes [17, 204, 209]. Nuclear scattering lengths for neutrons vary widely between isotopes of the same chemical species and thus provide contrast between them not seen by any other probe. Using this intrinsic property of neutrons to distinguish between isotopes of atoms has helped to study the self-diffusion in isotopic multilayers, using neutron reflectivity (NR) [17, 204, 209]. The difference between nuclear scattering lengths of stable isotopes of an element often provides large contrast in NR, whereas for XRR there will be no such contrast between isotopes because x-rays interact with the atomic charge cloud, which is the same from isotope to isotope [17]. A comparison of XRR and NR profiles for an isotopic $^{Natural}Ni/^{62}Ni$ multilayer suggested that the neutron sees large NSLD contrast for this system and gives Bragg peak corresponds to the bilayer periodicity and XRR treats this multilayer as a single layer due to zero ESLD contrast [17]. Several such isotopic multilayers have been mentioned in the review article [17], where NR was successfully applied to study self-diffusion due to inherent scattering length contrast. Using NR, Gupta *et al* [204] investigated the self-diffusion of Fe in a Fe isotope-based alloy, amorphous $^{Natural}Fe_{67}Zr_{33}/^{57}Fe_{67}Zr_{33}$ multilayers. Self-diffusion in the multilayer was measured after isothermal vacuum annealing below 573 K by monitoring the decay of the intensity of the Bragg peak of NR data. Diffusivity measured from NR data at different temperatures follows an Arrhenius-type behaviour with the activation energy E_a = 0.38 ± 0.05 eV. Using NR, Schmidt *et al* [209] determined nitrogen self-diffusivities ranging from 10^{-24} to 10^{-21} m^2 s^{-1} between 950 °C and 1250 °C in an isotopic N-based amorphous silicon nitride multilayer ($Si_3{}^{14}N_4/Si_3{}^{15}N_4$), which follow the Arrhenius law with an activation enthalpy of 3.6 eV in the relaxed state.

3.11 Proximity-driven magnetic order in topological insulators and interface magnetization in Weyl semimetal

Magnetic ordering in topological insulators (TIs) is crucial for breaking time-reversal symmetry and thereby opening a gap in the topological surface states. These properties are key to realizing useful topological properties such as the quantum anomalous Hall effect, the axion insulator state, and the topological magnetoelectric effect. Heterostructures of TIs with magnetic materials can be expected to yield interfaces with unique topological and magnetic phases [210, 211]. Li *et al* [210] observed a dramatic enhancement of proximity exchange coupling in the magnetic insulator (EuS)/TI ($Sb_{2-x}V_xTe_3$) hybrid heterostructure, where V doping is used to drive the TI (Sb_2Te_3) magnetic. Using PNR an increase of proximity magnetization per unit cell (u.c.) in TI, from 1.2 μ_B/u.c. to 2.7 μ_B/u.c. at $x = 0.1$ doping level was found. In another report, Katmis *et al* [211] successfully used the spin-dependent PNR technique to reveal the emergence of a ferromagnetic state in the top two quintuple layers (QL, where 1 QL ≈ 0.96 nm) of Bi_2Se_3 TI near the TI–FMI (Bi_2Se_3–EuS) interface. The PNR results also suggested that the ferromagnetic state in the TI persists up to temperatures larger than 300 K, far above the Curie temperature of ~17 K of bulk EuS. PNR measurements efficiently distinguished the magnetic TI surface states from the trivial bulk states and indicated that such effects originate through exchange interaction, without structural perturbation at the interface.

Magnetic Weyl semimetals are newly discovered quantum materials which are promising systems for use in spintronic applications. The cubic Heusler compound Co_2MnGa is a promising Weyl semimetal due to its inherent magnetic and topological properties [212]. Co_2MnGa shows a Curie temperature of T_c = 694 K and a saturation moment of M_s = 4.05 μ_B per formula unit [213]. Swekis *et al* [212] carried out specular PNR measurements and estimated the magnetization of the Co_2MnGa layer grown on the MgO substrate and observed a uniform depth profile of the film magnetization.

3.12 Control of local magnetization in isovalent oxide heterostructures by interface engineering

In the field of semiconductors, delta doping (δ doping) [214] was used to obtain high mobilities in two-dimensional electron gases, which has provided several fundamental discoveries and many novel applications. Delta doping was demonstrated in complex oxide heterostructures [194, 215] to enhance the magnetization by controlling the double exchange interaction locally using interface engineering. Santos *et al* [194] synthesized cation-ordered analogues of AFM $La_{0:5}Sr_{0:5}MnO_3$ (LSMO) by alternating single unit cell layers of $LaMnO_3$ (LMO) and $SrMnO_3$ (SMO) with atomic layer precision onto $SrTiO_3$ substrates using ozone-assisted molecular beam epitaxy and observed localized regions of enhanced FM exchange in an otherwise AFM structure. The high moment region occurs via a continuous variation of the canting angle that is governed by the spreading of charge near the delta-doped layer. Using specular PNR measurements, Santos *et al* [194] determined a length scale of 6 unit cells for the region of enhanced moments that is induced by the delta-doped layer. In another study, Moon *et al* [215] explored the 'δ-doping' strategy for the isovalent heterostructure of LSMO and $La_{0.5}Ca_{0.5}MnO_3$ (LCMO) by growing two unit cells of LSMO into thicker LCMO layers (20 unit cells) in a $[LSMO/LCMO]_5$ superlattices on an LSAT substrate. Combining specular PNR and TEM experimental measurements with the density functional theory, Moon *et al* [215] revealed enhancement in local magnetic exchange interactions and magnetization within the spatially confined regions of suppressed octahedral rotations due to structural δ doping. These studies further highlighted the strength of PNR techniques to investigate the emerging interfacial magnetism by controlled interface engineering of oxide heterostructures.

The different systems presented in this chapter highlight the use of specular XRR and PNR techniques as primary tools for revealing emerging interfacial magnetism at the interfaces of heterostructures of technological interest. These results vindicate the significant and extensive contribution of PNR in revealing emerging interface magnetism in heterostructures, suggesting that the interfacial magnetism at a depth can perhaps be best investigated using this vector magnetometric technique.

References

[1] Hellman F *et al* 2017 Interface-induced phenomena in magnetism *Rev. Mod. Phys.* **89** 025006

[2] Engel B N, England C D, Van Leeuwen R A, Wiedmann M H and Falco C M 1991 Interface magnetic anisotropy in epitaxial superlattices *Phys. Rev. Lett.* **67** 1910–3

[3] Carcia P F, Meinhaldt A D and Suna A 1985 Perpendicular magnetic anisotropy in Pd/Co thin film layered structures *Appl. Phys. Lett.* **47** 178–80

[4] den Broeder F J A, Hoving W and Bloemen P J H 1991 Magnetic anisotropy of multilayers *J. Magn. Magn. Mater.* **93** 562–70

[5] Chappert C, Dang K L, Beauvillain P, Hurdequint H and Renard D 1986 Ferromagnetic resonance studies of very thin cobalt films on a gold substrate *Phys. Rev.* B **34** 3192–7

[6] Stöhr J 1999 Exploring the microscopic origin of magnetic anisotropies with x-ray magnetic circular dichroism (XMCD) spectroscopy *J. Magn. Magn. Mater.* **200** 470–97

[7] Dzyaloshinsky I 1958 A thermodynamic theory of 'weak' ferromagnetism of antiferromagnetics *J. Phys. Chem. Solids* **4** 241–55

[8] Moriya T 1960 Anisotropic superexchange interaction and weak ferromagnetism *Phys. Rev.* **120** 91–8

[9] Meiklejohn W H and Bean C P 1956 New magnetic anisotropy *Phys. Rev.* **102** 1413–4

[10] Nogués J and Schuller I K 1999 Exchange bias *J. Magn. Magn. Mater.* **192** 203–32
[11] Baibich M N, Broto J M, Fert A, Van Dau F N, Petroff F, Etienne P, Creuzet G, Friederich A and Chazelas J 1988 Giant magnetoresistance of (001)Fe/(001)Cr magnetic superlattices *Phys. Rev. Lett.* **61** 2472–5
[12] Bland J A C and Heinrich B 1994 *Ultrathin Magnetic Structures* vol 1 (Berlin: Springer)
[13] Singh S, Basu S, Bhatt P and Poswal A K 2009 Kinetics of alloy formation at the interfaces in a Ni–Ti multilayer: x-ray and neutron reflectometry study *Phys. Rev.* B **79** 195435
[14] Singh S, Basu S, Gupta M, Majkrzak C F and Kienzle P A 2010 Growth kinetics of intermetallic alloy phase at the interfaces of a Ni/Al multilayer using polarized neutron and x-ray reflectometry *Phys. Rev.* B **81** 235413
[15] Swain M, Singh S, Basu S, Bhattacharya D and Gupta M 2014 Identification of a kinetic length scale which dictates alloy phase composition in Ni–Al interfaces on annealing at low temperatures *J. Appl. Phys.* **116** 222208
[16] Swain M, Singh S, Basu S and Gupta M 2013 Effect of interface morphology on intermetallics formation upon annealing of Al–Ni multilayer *J. Alloys Compd.* **576** 257–61
[17] Singh S, Swain M and Basu S 2018 Kinetics of interface alloy phase formation at nanometer length scale in ultra-thin films: x-ray and polarized neutron reflectometry *Prog. Mater. Sci.* **96** 1–50
[18] Fitzsimmons M R and Majkrzak C 2005 *Modern Techniques for Characterizing Magnetic Materials* (New York: Springer) pp 107–55
[19] Fitzsimmons M R *et al* 2011 Upper limit to magnetism in $LaAlO_3/SrTiO_3$ heterostructures *Phys. Rev. Lett.* **107** 217201
[20] Kikkawa J M and Awschalom D D 1999 Lateral drag of spin coherence in gallium arsenide *Nature* **397** 139–41
[21] Edmonds K W, Wang K Y, Campion R P, Neumann A C, Farley N R S, Gallagher B L and Foxon C T 2002 High-Curie-temperature $Ga_{1-x}Mn_xAs$ obtained by resistance-monitored annealing *Appl. Phys. Lett.* **81** 4991–3
[22] Krebs J J, Jonker B T and Prinz G A 1987 Properties of Fe single-crystal films grown on (100)GaAs by molecular-beam epitaxy *J. Appl. Phys.* **61** 2596–9
[23] Brillson L J 1982 The structure and properties of metal–semiconductor interfaces *Surf. Sci. Rep.* **2** 123–326
[24] Le Lay G 1983 Physics and electronics of the noble-metal/elemental-semiconductor interface formation: a status report *Surf. Sci.* **132** 169–204
[25] Perlov A, Popescu V and Ebert H 2002 *Ab initio* calculations of magnetoresistance and interlayer exchange constants of the ferromagnet–semiconductor layered systems Fe/Ge *J. Appl. Phys.* **91** 8798–800
[26] Gareev R R, Bürgler D E, Schreiber R, Braak H, Buchmeier M and Grünberg P A 2003 Antiferromagnetic interlayer exchange coupling across epitaxial, Ge-containing spacers *Appl. Phys. Lett.* **83** 1806–8
[27] Walser P, Schleberger M, Fuchs P and Landolt M 1998 Heat induced antiferromagnetic coupling in multilayers with Ge spacers *Phys. Rev. Lett.* **80** 2217–20
[28] Wolf S A *et al* 2001 Spintronics: a spin-based electronics vision for the future *Science* **294** 1488–95
[29] Datta S and Das B 1990 Electronic analog of the electro-optic modulator *Appl. Phys. Lett.* **56** 665–7

[30] Žutić I, Fabian J and Das Sarma S 2004 Spintronics: fundamentals and applications *Rev. Mod. Phys.* **76** 323–410
[31] Flatté M E and Vignale G 2001 Unipolar spin diodes and transistors *Appl. Phys. Lett.* **78** 1273–5
[32] Schmidt G, Ferrand D, Molenkamp L W, Filip A T and van Wees B J 2000 Fundamental obstacle for electrical spin injection from a ferromagnetic metal into a diffusive semiconductor *Phys. Rev.* B **62** R4790–3
[33] Gillingham D M, Tselepi M, Ionescu A, Steinmuller S J, Beere H E, Ritchie D A and Bland J A C 2007 Smoothening of ultrathin Fe films grown on GaAs(100) observed by scanning tunneling microscopy and Brillouin light scattering *Phys. Rev.* B **76** 214412
[34] Sze S M and Irvin J C 1968 Resistivity, mobility and impurity levels in GaAs, Ge, and Si at 300 K *Solid-State Electron.* **11** 599–602
[35] Jaafar R, Berling D, Sébilleau D and Garreau G 2010 Epitaxial Fe–Ge thin films on Ge (111): morphology, structure, and magnetic properties versus stoichiometry *Phys. Rev.* B **81** 155423
[36] Lin Y-C, Chen Y, Chen R, Ghosh K, Xiong Q and Huang Y 2012 Crystallinity control of ferromagnetic contacts in stressed nanowire templates and the magnetic domain anisotropy *Nano Lett.* **12** 4341–8
[37] Cabria I, Perlov A Y and Ebert H 2001 Magnetization profile and magnetocrystalline anisotropy of ferromagnet–semiconductor heterostructure systems *Phys. Rev.* B **63** 104424
[38] Butler W H, Zhang X G, Wang X, van Ek J and MacLaren J M 1997 Electronic structure of FM|semiconductor|FM spin tunneling structures *J. Appl. Phys.* **81** 5518–20
[39] Schleberger M, Walser P, Hunziker M and Landolt M 1999 Amorphous Fe–Si and Fe–Ge nanostructures quantitatively analyzed by x-ray-photoelectron spectroscopy *Phys. Rev.* B **60** 14360–5
[40] Singh S, Basu S, Vedpathak M, Kodama R H, Chitra R and Goud Y 2005 Ultrathin multilayer of Fe and Ge: structure and magnetic properties *Appl. Surf. Sci.* **240** 251–9
[41] Singh S, Basu S, Gupta M, Vedpathakz M and Kodama R H 2007 Investigation of interface magnetic moment of Fe/Ge multilayer: a neutron reflectivity study *J. Appl. Phys.* **101** 033913
[42] Singh S, Basu S, Bhattacharya D and Poswal A K 2010 Physical and magnetic roughness at metal–semiconductor interface using x-ray and neutron reflectometry *J. Appl. Phys.* **107** 123903
[43] Singh S, Bhatt H, Kumar Y, Prajapat C L, Mishra A, Bedanta S and Basu S 2021 Interface-driven static and dynamic magnetic properties of ultrathin Fe/Ge multilayers *Appl. Surf. Sci.* **570** 151193
[44] Majkrzak C F 1996 Neutron scattering studies of magnetic thin films and multilayers *Physica* B **221** 342–56
[45] Swain M, Singh S, Bhattacharya D, Singh A, Tokas R B, Prajapat C L and Basu S 2015 Micro-structural characterization of low resistive metallic Ni germanide growth on annealing of Ni–Ge multilayer *AIP Adv.* **5** 077129
[46] Grünberg P, Schreiber R, Pang Y, Brodsky M B and Sowers H 1986 Layered magnetic structures: evidence for antiferromagnetic coupling of Fe layers across Cr interlayers *Phys. Rev. Lett.* **57** 2442–5
[47] Duine R A, Lee K-J, Parkin S S P and Stiles M D 2018 Synthetic antiferromagnetic spintronics *Nat. Phys.* **14** 217–9

[48] Binasch G, Grünberg P, Saurenbach F and Zinn W 1989 Enhanced magnetoresistance in layered magnetic structures with antiferromagnetic interlayer exchange *Phys. Rev.* B **39** 4828–30
[49] Parkin S S P, More N and Roche K P 1990 Oscillations in exchange coupling and magnetoresistance in metallic superlattice structures: Co/Ru, Co/Cr, and Fe/Cr *Phys. Rev. Lett.* **64** 2304–7
[50] Parkin S S P 1991 Systematic variation of the strength and oscillation period of indirect magnetic exchange coupling through the 3*d*, 4*d*, and 5*d* transition metals *Phys. Rev. Lett.* **67** 3598–601
[51] Baltensperger W and Helman J S 1990 Ruderman–Kittel coupling between ferromagnets separated by a nonmagnetic layer *Appl. Phys. Lett.* **57** 2954–5
[52] Bruno P and Chappert C 1991 Oscillatory coupling between ferromagnetic layers separated by a nonmagnetic metal spacer *Phys. Rev. Lett.* **67** 1602–5
[53] Coehoorn R 1991 Period of oscillatory exchange interactions in Co/Cu and Fe/Cu multilayer systems *Phys. Rev.* B **44** 9331–7
[54] Bruno P and Chappert C 1992 Ruderman–Kittel theory of oscillatory interlayer exchange coupling *Phys. Rev.* B **46** 261–70
[55] Bruno P 1993 Interlayer exchange coupling: a unified physical picture *J. Magn. Magn. Mater.* **121** 248–52
[56] Edwards D M, Mathon J, Muniz R B and Phan M S 1991 Oscillations of the exchange in magnetic multilayers as an analog of de Haas–van Alphen effect *Phys. Rev. Lett.* **67** 493–6
[57] Rührig M, Schäfer R, Hubert A, Mosler R, Wolf J A, Demokritov S and Grünberg P 1991 Domain observations on Fe–Cr–Fe layered structures. Evidence for a biquadratic coupling effect *Phys. Status Solidi* a **125** 635–56
[58] Fuß A, Demokritov S, Grünberg P and Zinn W 1992 Short- and long period oscillations in the exchange coupling of Fe across epitaxially grown Al- and Au-interlayers *J. Magn. Magn. Mater.* **103** L221–7
[59] Gutierrez C J, Krebs J J, Filipkowski M E and Prinz G A 1992 Strong temperature dependence of the 90° coupling in Fe/Al/Fe(001) magnetic trilayers *J. Magn. Magn. Mater.* **116** L30510
[60] Slonczewski J C 1993 Origin of biquadratic exchange in magnetic multilayers (invited) *J. Appl. Phys.* **73** 5957–62
[61] Filipkowski M E, Gutierrez C J, Krebs J J and Prinz G A 1993 Temperature dependence of the 90° coupling in Fe/Al/Fe(001) magnetic trilayers *J. Appl. Phys.* **73** 5963–5
[62] Celinski Z, Heinrich B and Cochran J F 1993 Analysis of bilinear and biquadratic exchange coupling in Fe/Ag/Fe(001) trilayers *J. Appl. Phys.* **73** 5966–8
[63] Unguris J, Celotta R J and Pierce D T 1991 Observation of two different oscillation periods in the exchange coupling of Fe/Cr/Fe(100) *Phys. Rev. Lett.* **67** 140–3
[64] Unguris J, Celotta R J, Pierce D T and Stroscio J A 1993 Biquadratic exchange coupling in Fe/Cr/Fe(100) and Fe/Ag/Fe(100) thin film sandwiches (abstract) *J. Appl. Phys.* **73** 5984
[65] Heinrich B and Cochran J F 1993 Ultrathin metallic magnetic films: magnetic anisotropies and exchange interactions *Adv. Phys.* **42** 523–639
[66] Schreiber F, Soliman A, Bödeker P, Meckenstock R, Bröhl K, Pelzl J and Garifullin I A 1994 Anisotropy studies of molecular-beam-epitaxy-grown Co(111) thin films by ferromagnetic resonance *J. Appl. Phys.* **75** 6492–4

[67] Demokritov S, Wolf J A and Grünberg P 1991 Evidence for oscillations in the interlayer coupling of Fe films across Cr films from spin waves and *M*(*H*) curves *Europhys. Lett.* **15** 881–6
[68] Langridge S, Schmalian J, Marrows C H, Dekadjevi D T and Hickey B J 2000 Quantification of magnetic domain disorder and correlations in antiferromagnetically coupled multilayers by neutron reflectometry *Phys. Rev. Lett.* **85** 4964–7
[69] Holloway H and Kubinski D J 1998 Magnetoresistive Co/Cu multilayers: hysteresis, polycrystallinity, and irreversible changes on magnetization *J. Appl. Phys.* **83** 2705–14
[70] Wang Y, Levy P M and Fry J L 1990 Interlayer magnetic coupling in Fe/Cr multilayered structures *Phys. Rev. Lett.* **65** 2732–5
[71] Paul A, Kentzinger E, Rücker U, Bürgler D E and Brückel T 2006 Field-dependent magnetic domain structure in antiferromagnetically coupled multilayers by polarized neutron scattering *Phys. Rev.* B **73** 094441
[72] Cebollada A, Martnez J L, Gallego J M, de Miguel J J, Miranda R, Ferrer S, Batallán F, Fillion G and Rebouillat J P 1989 Antiferromagnetic ordering in Co–Cu single-crystal superlattices *Phys. Rev.* B **39** 9726–9
[73] Schreyer A, Bröhl K, Ankner J F, Majkrzak C F, Zeidler T, Bödeker P, Metoki N and Zabel H 1993 Oscillatory exchange coupling in Co/Cu(111) superlattices *Phys. Rev.* B **47** 15334–7
[74] Borchers J A, Dura J A, Unguris J, Tulchinsky D, Kelley M H, Majkrzak C F, Hsu S Y, Loloee R, Pratt W P and Bass J 1999 Observation of antiparallel magnetic order in weakly coupled Co/Cu multilayers *Phys. Rev. Lett.* **82** 2796–9
[75] Schreyer A, Ankner J F, Zeidler T, Zabel H, Schäfer M, Wolf J A, Grünberg P and Majkrzak C F 1995 Noncollinear and collinear magnetic structures in exchange coupled Fe/Cr(001) superlattices *Phys. Rev.* B **52** 16066–85
[76] Ustinov V V, Bebenin N G, Romashev L N, Minin V I, Milyaev M A, Del A R and Semerikov A V 1996 Magnetoresistance and magnetization of Fe/Cr(001) superlattices with noncollinear magnetic ordering *Phys. Rev.* B **54** 15958–66
[77] Lauter-Pasyuk V, Lauter H J, Toperverg B P, Romashev L and Ustinov V 2002 Transverse and lateral structure of the spin-flop phase in Fe/Cr antiferromagnetic superlattices *Phys. Rev. Lett.* **89** 167203
[78] Andreeva M A and Lindgren B 2005 Nuclear resonant spectroscopy at Bragg reflections from periodic multilayers: basic effects and applications *Phys. Rev.* B **72** 125422
[79] Andreeva M A *et al* 2018 Double-spiral magnetic structure of the Fe/Cr multilayer revealed by nuclear resonance reflectivity *Phys. Rev.* B **97** 024417
[80] Magnus F, Warnatz T, Palsson G K, Devishvili A, Ukleev V, Palisaitis J, Persson P O Å and Hjörvarsson B 2018 Sequential magnetic switching in Fe/MgO(001) superlattices *Phys. Rev.* B **97** 174424
[81] Ueda S, Iwasaki Y, Uehara Y and Ushioda S 2011 Changes of interlayer exchange coupling in Fe/Si/Fe trilayer structure by photo-irradiation *Phys. Rev.* B **83** 144424
[82] Sakuraba Y, Izumi K, Bosu S, Saito K and Takanashi K 2011 Temperature dependence of spin-dependent transport properties of Co_2MnSi-based current-perpendicular-to-plane magnetoresistive devices *J. Phys.* D **44** 064009
[83] Chung J H, Chung S J, Lee S, Kirby B J, Borchers J A, Cho Y J, Liu X and Furdyna J K 2008 Carrier-mediated antiferromagnetic interlayer exchange coupling in diluted magnetic semiconductor multilayers $Ga_{1-x}Mn_xAs/GaAs{:}Be$ *Phys. Rev. Lett.* **101** 237202

[84] Saerbeck T *et al* 2011 Spatial fluctuations of loose spin coupling in CuMn/Co multilayers *Phys. Rev. Lett.* **107** 127201
[85] Yu X Z, Onose Y, Kanazawa N, Park J H, Han J H, Matsui Y, Nagaosa N and Tokura Y 2010 Real-space observation of a two-dimensional skyrmion crystal *Nature* **465** 901–4
[86] Yang H X, Chshiev M, Dieny B, Lee J H, Manchon A and Shin K H 2011 First-principles investigation of the very large perpendicular magnetic anisotropy at Fe|MgO and Co|MgO interfaces *Phys. Rev.* B **84** 054401
[87] Bode M, Heide M, von Bergmann K, Ferriani P, Heinze S, Bihlmayer G, Kubetzka A, Pietzsch O, Blügel S and Wiesendanger R 2007 Chiral magnetic order at surfaces driven by inversion asymmetry *Nature* **447** 190–3
[88] Manchon A, Koo H C, Nitta J, Frolov S M and Duine R A 2015 New perspectives for Rashba spin–orbit coupling *Nat. Mater.* **14** 871–82
[89] Jungwirth T, Wunderlich J and Olejník K 2012 Spin Hall effect devices *Nat. Mater.* **11** 382–90
[90] Sinova J, Valenzuela S O, Wunderlich J, Back C H and Jungwirth T 2015 Spin Hall effects *Rev. Mod. Phys.* **87** 1213–60
[91] Ralph D C and Stiles M D 2008 Spin transfer torques *J. Magn. Magn. Mater.* **320** 1190–216
[92] Brataas A, Kent A D and Ohno H 2012 Current-induced torques in magnetic materials *Nat. Mater.* **11** 372–81
[93] Fert A and Levy P M 1980 Role of anisotropic exchange interactions in determining the properties of spin-glasses *Phys. Rev. Lett.* **44** 1538–41
[94] Romming N, Hanneken C, Menzel M, Bickel J E, Wolter B, von Bergmann K, Kubetzka A and Wiesendanger R 2013 Writing and deleting single magnetic skyrmions *Science* **341** 636–9
[95] Chen G, Ma T, N'Diaye A T, Kwon H, Won C, Wu Y and Schmid A K 2013 Tailoring the chirality of magnetic domain walls by interface engineering *Nat. Commun.* **4** 2671
[96] Sunko V *et al* 2017 Maximal Rashba-like spin splitting via kinetic-energy-coupled inversion-symmetry breaking *Nature* **549** 492–6
[97] Hirsch J E 1999 Spin Hall effect *Phys. Rev. Lett.* **83** 1834–7
[98] Slonczewski J C 1996 Current-driven excitation of magnetic multilayers *J. Magn. Magn. Mater.* **159** L1–7
[99] Berger L 1996 Emission of spin waves by a magnetic multilayer traversed by a current *Phys. Rev.* B **54** 9353–8
[100] Hase T P A *et al* 2014 Proximity effects on dimensionality and magnetic ordering in Pd/Fe/Pd trialyers *Phys. Rev.* B **90** 104403
[101] Mayr S, Ye J, Stahn J, Knoblich B, Klein O, Gilbert D A, Albrecht M, Paul A, Böni P and Kreuzpaintner W 2020 Indications for Dzyaloshinskii–Moriya interaction at the Pd/Fe interface studied by *in situ* polarized neutron reflectometry *Phys. Rev.* B **101** 024404
[102] Yang H, Thiaville A, Rohart S, Fert A and Chshiev M 2015 Anatomy of Dzyaloshinskii–Moriya interaction at Co/Pt interfaces *Phys. Rev. Lett.* **115** 267210
[103] Judy J H 2005 Advancements in PMR thin-film media *J. Magn. Magn. Mater.* **287** 16–26
[104] Romera M, Muñoz M, Maicas M, Michalik J M, de Teresa J M, Magén C and Prieto J L 2011 Enhanced exchange and reduced magnetization of Gd in an Fe/Gd/Fe trilayer *Phys. Rev.* B **84** 094456
[105] Haskel D, Srajer G, Lang J C, Pollmann J, Nelson C S, Jiang J S and Bader S D 2001 Enhanced interfacial magnetic coupling of Gd/Fe multilayers *Phys. Rev. Lett.* **87** 207201
[106] Camley R E and Tilley D R 1988 Phase transitions in magnetic superlattices *Phys. Rev.* B **37** 3413–21

[107] Kim K-J *et al* 2017 Fast domain wall motion in the vicinity of the angular momentum compensation temperature of ferrimagnets *Nat. Mater.* **16** 1187–92
[108] Siddiqui S A, Han J, Finley J T, Ross C A and Liu L 2018 Current-induced domain wall motion in a compensated ferrimagnet *Phys. Rev. Lett.* **121** 057701
[109] Bläsing R, Ma T, Yang S-H, Garg C, Dejene F K, N'Diaye A T, Chen G, Liu K and Parkin S S P 2018 Exchange coupling torque in ferrimagnetic Co/Gd bilayer maximized near angular momentum compensation temperature *Nat. Commun.* **9** 4984
[110] Dzemiantsova L V, Meier G and Röhlsberger R 2015 Stabilization of magnetic helix in exchange-coupled thin films *Sci. Rep.* **5** 16153
[111] Fust S, Mukherjee S, Paul N, Stahn J, Kreuzpaintner W, Böni P and Paul A 2016 Realizing topological stability of magnetic helices in exchange-coupled multilayers for all-spin-based system *Sci. Rep.* **6** 33986
[112] Singh S *et al* 2019 Antisymmetric magnetoresistance and helical magnetic structure in a compensated Gd/Co multilayer *Phys. Rev.* B **100** 140405
[113] Eschrig M 2010 Spin-polarized supercurrents for spintronics *Phys. Today* **64** 43–9
[114] Bergeret F S, Volkov A F and Efetov K B 2005 Odd triplet superconductivity and related phenomena in superconductor–ferromagnet structures *Rev. Mod. Phys.* **77** 1321–73
[115] Mangin S *et al* 2014 Engineered materials for all-optical helicity-dependent magnetic switching *Nat. Mater.* **13** 286–92
[116] LePage J G and Camley R E 1990 Surface phase transitions and spin-wave modes in semi-infinite magnetic superlattices with antiferromagnetic interfacial coupling *Phys. Rev. Lett.* **65** 1152–5
[117] Sajieddine M, Bauer P, Cherifi K, Dufour C, Marchal G and Camley R E 1994 Experimental and theoretical spin configurations in Fe/Gd multilayers *Phys. Rev.* B **49** 8815–20
[118] Haskel D, Srajer G, Choi Y, Lee D R, Lang J C, Meersschaut J, Jiang J S and Bader S D 2003 Nature of inhomogeneous magnetic state in artificial Fe/Gd ferrimagnetic multilayers *Phys. Rev.* B **67** 180406
[119] Kravtsov E, Haskel D, te Velthuis S G E, Jiang J S and Kirby B J 2009 Complementary polarized neutron and resonant x-ray magnetic reflectometry measurements in Fe/Gd heterostructures: case of inhomogeneous intralayer magnetic structure *Phys. Rev.* B **79** 134438
[120] Prieto J L, van Aken B B, Burnell G, Bell C, Evetts J E, Mathur N and Blamire M G 2004 Transport properties of sharp antiferromagnetic boundaries in Gd/Fe multilayers *Phys. Rev.* B **69** 054436
[121] Hahn W, Loewenhaupt M, Huang Y Y, Felcher G P and Parkin S S P 1995 Experimental determination of the magnetic phase diagram of Gd/Fe multilayers *Phys. Rev.* B **52** 16041–8
[122] Ishimatsu N, Hashizume H, Hamada S, Hosoito N, Nelson C S, Venkataraman C T, Srajer G and Lang J C 1999 Magnetic structure of Fe/Gd multilayers determined by resonant x-ray magnetic scattering *Phys. Rev.* B **60** 9596–606
[123] Choi Y, Haskel D, Cady A, Lang J C, Lee D R, Srajer G, Jiang J S and Bader S D 2006 Twisted magnetization states near the compensation temperature of Fe/Gd multilayers: anisotropy and surface-termination effects *Phys. Rev.* B **73** 174401
[124] Montoya S A *et al* 2017 Tailoring magnetic energies to form dipole skyrmions and skyrmion lattices *Phys. Rev.* B **95** 024415
[125] Desautels R D, DeBeer-Schmitt L, Montoya S A, Borchers J A, Je S-G, Tang N, Im M-Y, Fitzsimmons M R, Fullerton E E and Gilbert D A 2019 Realization of ordered magnetic skyrmions in thin films at ambient conditions *Phys. Rev. Mater.* **3** 104406

[126] Camley R E and Stamps R L 1993 Magnetic multilayers: spin configurations, excitations and giant magnetoresistance *J. Phys. Condens. Matter* **5** 3727–86

[127] Schlage K and Röhlsberger R 2013 Nuclear resonant scattering of synchrotron radiation: applications in magnetism of layered structures *J. Electron. Spectrosc. Relat. Phenom.* **189** 187–95

[128] Fink J, Schierle E, Weschke E and Geck J 2013 Resonant elastic soft x-ray scattering *Rep. Prog. Phys.* **76** 056502

[129] Basha M A *et al* 2020 Helical magnetic structure and exchange bias across the compensation temperature of Gd/Co multilayers *J. Appl. Phys.* **128** 103901

[130] Basha M A, Prajapat C L, Gupta M, Bhatt H, Kumar Y, Ghosh S K, Karki V, Basu S and Singh S 2018 Interface induced magnetic properties of Gd/Co heterostructures *Phys. Chem. Chem. Phys.* **20** 21580–9

[131] Higgs T D C, Bonetti S, Ohldag H, Banerjee N, Wang X L, Rosenberg A J, Cai Z, Zhao J H, Moler K A and Robinson J W A 2016 Magnetic coupling at rare earth ferromagnet/transition metal ferromagnet interfaces: a comprehensive study of Gd/Ni *Sci. Rep.* **6** 30092

[132] Lapa P N, Ding J, Pearson J E, Novosad V, Jiang J S and Hoffmann A 2017 Magnetization reversal in Py/Gd heterostructures *Phys. Rev.* B **96** 024418

[133] Tappert J *et al* 2000 Fe spin structure in Tb/Fe multilayers *Philos. Mag.* B **80** 257–74

[134] Gupta A, Paul A, Gupta R, Avasthi D K and Principi G 1998 The effect of swift heavy ion irradiation on perpendicular magnetic anisotropy in Fe–Tb multilayers *J. Phys. Condens. Matter* **10** 9669–80

[135] Takanod K, Ikeuchi K, Sakurai H, Oike H and Itoh F 2004 Magnetic twisted state of Fe/Tb multilayers *J. Phys. Chem. Solids* **65** 1985–92

[136] Paul A, Mukherjee S, Kreuzpaintner W and Böni P 2014 Exchange-bias-like coupling in a ferrimagnetic Fe/Tb multilayer *Phys. Rev.* B **89** 144415

[137] Mukherjee S, Kreuzpaintner W, Stahn J, Zheng J-G, Bauer A, Böni P and Paul A 2015 Exchange-bias-like coupling in a Cu-diluted-Fe/Tb multilayer *Phys. Rev.* B **91** 104419

[138] Stanciu C D, Hansteen F, Kimel A V, Kirilyuk A, Tsukamoto A, Itoh A and Rasing T 2007 All-optical magnetic recording with circularly polarized light *Phys. Rev. Lett.* **99** 047601

[139] Je S-G *et al* 2018 Spin–orbit torque-induced switching in ferrimagnetic alloys: experiments and modeling *Appl. Phys. Lett.* **112** 062401

[140] Chen J-Y, He L, Wang J-P and Li M 2017 All-optical switching of magnetic tunnel junctions with single subpicosecond laser pulses *Phys. Rev. Appl.* **7** 021001

[141] Yu J *et al* 2019 Long spin coherence length and bulk-like spin–orbit torque in ferrimagnetic multilayers *Nat. Mater.* **18** 29–34

[142] Majkrzak C F, Cable J W, Kwo J, Hong M, McWhan D B, Yafet Y, Waszczak J V and Vettier C 1986 Observation of a magnetic antiphase domain structure with long-range order in a synthetic Gd–Y superlattice *Phys. Rev. Lett.* **56** 2700–3

[143] Salamon M B, Sinha S, Rhyne J J, Cunningham J E, Erwin R W, Borchers J and Flynn C P 1986 Long-range incommensurate magnetic order in a Dy–Y multilayer *Phys. Rev. Lett.* **56** 259–62

[144] Grigoriev S V, Chetverikov Y O, Lott D and Schreyer A 2008 Field induced chirality in the helix structure of Dy/Y multilayer films and experimental evidence for Dzyaloshinskii–Moriya interaction on the interfaces *Phys. Rev. Lett.* **100** 197203

[145] Antropov N O, Kravtsov E A, Khaidukov Y N, Ryabukhina M V, Proglyado V V, Weschke O and Ustinov V V 2018 Coherent fan magnetic structure in Dy/Gd superlattices *JETP Lett.* **108** 341–5

[146] Paul A 2016 Stiffness in vortex-like structures due to chirality-domains within a coupled helical rare-earth superlattice *Sci. Rep.* **6** 19315

[147] Ye J, Baldauf T, Mattauch S, Paul N and Paul A 2019 Topologically stable helices in exchange coupled rare-earth/rare-earth multilayer with superspin-glass like ordering *Commun. Phys.* **2** 114

[148] Fitzsimmons M R, Leighton C, Hoffmann A, Yashar P C, Nogués J, Liu K, Majkrzak C F, Dura J A, Fritzsche H and Schuller I K 2001 Influence of interfacial disorder and temperature on magnetization reversal in exchange-coupled bilayers *Phys. Rev.* B **64** 104415

[149] Fitzsimmons M R *et al* 2002 Influence of in-plane crystalline quality of an antiferromagnet on perpendicular exchange coupling and exchange bias *Phys. Rev.* B **65** 134436

[150] Moran T J and Schuller I K 1996 Effects of cooling field strength on exchange anisotropy at permalloy/CoO interfaces *J. Appl. Phys.* **79** 5109–11

[151] Nogués J, Lederman D, Moran T J and Schuller I K 1996 Positive exchange bias in FeF_2–Fe bilayers *Phys. Rev. Lett.* **76** 4624–7

[152] Nogués J, Moran T J, Lederman D, Schuller I K and Rao K V 1999 Role of interfacial structure on exchange-biased FeF_2–Fe *Phys. Rev.* B **59** 6984–93

[153] Paul A, Schneider C M and Stahn J 2007 Reversal mechanism and suppression of training in an exchange-coupled system *Phys. Rev.* B **76** 184424

[154] Roy S *et al* 2005 Depth profile of uncompensated spins in an exchange bias system *Phys. Rev. Lett.* **95** 047201

[155] Fitzsimmons M R, Kirby B J, Roy S, Li Z-P, Roshchin I V, Sinha S K and Schuller I K 2007 Pinned magnetization in the antiferromagnet and ferromagnet of an exchange bias system *Phys. Rev.* B **75** 214412

[156] Ball A R, Leenaers A J G, van der Zaag P J, Shaw K A, Singer B, Lind D M, Fredrikze H and Rekveldt M T 1996 Polarized neutron reflectometry study of an exchange biased Fe_3O_4/NiO multilayer *Appl. Phys. Lett.* **69** 583–5

[157] Dufour C, Fitzsimmons M R, Borchers J A, Laver M, Krycka K L, Dumesnil K, Watson S M, Chen W C, Won J and Singh S 2011 Nanometer-size magnetic domains and coherent magnetization reversal in a giant exchange-bias system *Phys. Rev.* B **84** 064420

[158] Radu F, Vorobiev A, Major J, Humblot H, Westerholt K and Zabel H 2003 Spin-resolved off-specular neutron scattering from magnetic domain walls using the polarized ^{3}He gas spin filter *Physica* B **335** 63–7

[159] Prajapat C L *et al* 2020 Interface-induced magnetization and exchange bias in LSMO/BFO multiferroic heterostructures *ACS Appl. Electron. Mater.* **2** 2629–37

[160] Bhatt H, Kumar Y, Prajapat C L, Kinane C J, Caruana A, Langridge S, Basu S and Singh S 2020 Emergent interfacial ferromagnetism and exchange bias effect in paramagnetic/ferromagnetic oxide heterostructures *Adv. Mater. Interfaces* **7** 2001172

[161] He C *et al* 2012 Interfacial ferromagnetism and exchange bias in $CaRuO_3/CaMnO_3$ superlattices *Phys. Rev. Lett.* **109** 197202

[162] Basha M A, Bhatt H, Kumar Y, Prajapat C L, Gupta M, Karki V, Basu S and Singh S 2020 Annealing driven positive and negative exchange bias in Fe–Cu–Pt heterostructures at room temperature *J. Alloys Compd.* **815** 152640

[163] Estrader M *et al* 2013 Robust antiferromagnetic coupling in hard-soft bi-magnetic core/shell nanoparticles *Nat. Commun.* **4** 2960

[164] Sharma A, Tripathi J, Tripathi S, Kumar Y, Ugochukwu K C, Kumar D, Gupta M and Chaudhary R J 2020 Exchange bias in Co/CoO thin films deposited onto self-assembled nanosphere arrays *J. Magn. Magn. Mater.* **510** 166599

[165] Radu F, Etzkorn M, Leiner V, Schmitte T, Schreyer A, Westerholt K and Zabel H 2002 Polarised neutron reflectometry study of Co/CoO exchange-biased multilayers *Appl. Phys.* A **74** s1570–2

[166] Raghavan L, Ranjith K M, Baenitz M and Kanjilal D 2020 Control of exchange bias by modifying the antiferromagnet and ferromagnet–antiferromagnet interface *Europhys. Lett.* **130** 67006

[167] Anyfantis D I, Sarigiannidou E, Rapenne L, Stamatelatos A, Ntemogiannis D, Kapaklis V and Poulopoulos P 2019 Unexpected development of perpendicular magnetic anisotropy in Ni/NiO multilayers after mild thermal annealing *IEEE Magn. Lett.* **10** 1–5

[168] Zhang Y-J, Chen J-H, Li L-L, Ma J, Nan C-W and Lin Y-H 2017 Ferroelectric strain modulation of antiferromagnetic moments in Ni/NiO ferromagnet/antiferromagnet heterostructures *Phys. Rev.* B **95** 174420

[169] Lazar L, Jiang J S, Felcher G P, Inomata A and Bader S D 2001 Oscillatory exchange bias in Fe/Cr double superlattices *J. Magn. Magn. Mater.* **223** 299–303

[170] Steadman P, Ali M, Hindmarch A T, Marrows C H, Hickey B J, Langridge S, Dalgliesh R M and Foster S 2002 Exchange bias in spin-engineered double superlattices *Phys. Rev. Lett.* **89** 077201

[171] Marrows C H, Langridge S, Ali M, Hindmarch A T, Dekadjevi D T, Foster S and Hickey B J 2002 Mapping domain disorder in exchange-biased magnetic multilayers *Phys. Rev.* B **66** 024437

[172] Hellman F, van Dover R B and Gyorgy E M 1987 Unexpected unidirectional anisotropy in amorphous Tb–Fe/Ni–Fe–Mo bilayer films *Appl. Phys. Lett.* **50** 296–8

[173] Altuncevahir B and Koymen A R 2003 Positive and negative exchange bias in CoNi/Gd/CoNi trilayers and CoNi/Gd bilayers *J. Magn. Magn. Mater.* **261** 424–32

[174] Hauet T, Borchers J A, Mangin P, Henry Y and Mangin S 2006 Training effect in an exchange bias system: the role of interfacial domain walls *Phys. Rev. Lett.* **96** 067207

[175] Singh S, Basha M A, Bhatt H, Kumar Y and Gupta M 2022 Interface morphology driven exchange interaction and magnetization reversal in a Gd/Co multilayer *Phys. Chem. Chem. Phys.* **24** 6580–9

[176] Satapathy D K *et al* 2012 Magnetic proximity effect in $YBa_2Cu_3O_7/La_{2/3}Ca_{1/3}MnO_3$ and $YBa_2Cu_3O_7/LaMnO_{3+\delta}$ superlattices *Phys. Rev. Lett.* **108** 197201

[177] Buzdin A I 2005 Proximity effects in superconductor–ferromagnet heterostructures *Rev. Mod. Phys.* **77** 935–76

[178] Chakhalian J *et al* 2006 Magnetism at the interface between ferromagnetic and superconducting oxides *Nat. Phys.* **2** 244–8

[179] Hoppler J *et al* 2009 Giant superconductivity-induced modulation of the ferromagnetic magnetization in a cuprate–manganite superlattice *Nat. Mater.* **8** 315–9

[180] Prajapat C L, Singh S, Paul A, Bhattacharya D, Singh M R, Mattauch S, Ravikumar G and Basu S 2016 Superconductivity-induced magnetization depletion in a ferromagnet through an insulator in a ferromagnet–insulator–superconductor hybrid oxide heterostructure *Nanoscale* **8** 10188–97

[181] Prajapat C L, Singh S, Bhattacharya D, Ravikumar G, Basu S, Mattauch S, Zheng J-G, Aoki T and Paul A 2018 Proximity effects across oxide-interfaces of superconductor–insulator–ferromagnet hybrid heterostructure *Sci. Rep.* **8** 3732

[182] Paull O H C, Pan A V, Causer G L, Fedoseev S A, Jones A, Liu X, Rosenfeld A and Klose F 2018 Field dependence of the ferromagnetic/superconducting proximity effect in a YBCO/STO/LCMO multilayer *Nanoscale* **10** 18995–9003

[183] Singh S, Bhatt H, Kumar Y, Prajapat C L, Satpati B, Kinane C J, Langridge S, Ravikumar G and Basu S 2020 Superconductivity-driven negative interfacial magnetization in $YBa_2Cu_3O_{7-\delta}/SrTiO_3/La_{0.67}Sr_{0.33}MnO_3$ heterostructures *Appl. Phys. Lett.* **116** 022406

[184] Kumar Y, Bhatt H, Prajapat C L, Singh A P, Singh F, Kinane C J, Langridge S, Basu S and Singh S 2021 Suppression of the superconducting proximity effect in ferromagnetic-superconducting oxide heterostructures with ion-irradiation *J. Appl. Phys.* **129** 163902

[185] Bhatt H, Kumar Y, Prajapat C L, Kinane C J, Caruana A, Langridge S, Basu S and Singh S 2022 Correlation of magnetic and superconducting properties with the strength of the magnetic proximity effect in $La_{0.67}Sr_{0.33}MnO_3/SrTiO_3/YBa_2Cu_3O_{7-\delta}$ heterostructures *ACS Appl. Mater. Interfaces* **14** 8565–74

[186] Hwang H Y, Iwasa Y, Kawasaki M, Keimer B, Nagaosa N and Tokura Y 2012 Emergent phenomena at oxide interfaces *Nat. Mater.* **11** 103–13

[187] Ahn K H, Lookman T and Bishop A R 2004 Strain-induced metal–insulator phase coexistence in perovskite manganites *Nature* **428** 401–4

[188] Adamo C *et al* 2009 Effect of biaxial strain on the electrical and magnetic properties of (001) $La_{0.7}Sr_{0.3}MnO_3$ thin films *Appl. Phys. Lett.* **95** 112504

[189] Singh S, Fitzsimmons M R, Lookman T, Jeen H, Biswas A, Roldan M A and Varela M 2012 Role of elastic bending stress on magnetism of a manganite thin film studied by polarized neutron reflectometry *Phys. Rev.* B **85** 214440

[190] Wang Q, Chen A P, Guo E J, Roldan M A, Jia Q X and Fitzsimmons M R 2018 Upper limit for the effect of elastic bending stress on the saturation magnetization of $La_{0.8}Sr_{0.2}MnO_3$ *Phys. Rev.* B **97** 014437

[191] Ueda K, Tabata H and Kawai T 2001 Control of magnetic properties in $LaCrO_3$–$LaFeO_3$ artificial superlattices *J. Appl. Phys.* **89** 2847–51

[192] Salvador P A, Haghiri-Gosnet A M, Mercey B, Hervieu M and Raveau B 1999 Growth and magnetoresistive properties of $(LaMnO_3)_m(SrMnO_3)_n$ superlattices *Appl. Phys. Lett.* **75** 2638–40

[193] Lin C, Okamoto S and Millis A J 2006 Dynamical mean-field study of model double-exchange superlattices *Phys. Rev.* B **73** 041104

[194] Santos T S, Kirby B J, Kumar S, May S J, Borchers J A, Maranville B B, Zarestky J, te Velthuis S G E, van den Brink J and Bhattacharya A 2011 Delta doping of ferromagnetism in antiferromagnetic manganite superlattices *Phys. Rev. Lett.* **107** 167202

[195] Grutter A J *et al* 2013 Interfacial ferromagnetism in $LaNiO_3/CaMnO_3$ superlattices *Phys. Rev. Lett.* **111** 087202

[196] Nanda B R K, Satpathy S and Springborg M S 2007 Electron leakage and double-exchange ferromagnetism at the interface between a metal and an antiferromagnetic insulator: $CaRuO_3/CaMnO_3$ *Phys. Rev. Lett.* **98** 216804

[197] Freeland J W *et al* 2010 Charge transport and magnetization profile at the interface between the correlated metal $CaRuO_3$ and the antiferromagnetic insulator $CaMnO_3$ *Phys. Rev.* B **81** 094414

[198] Hoffman J, Tung I C, Nelson-Cheeseman B B, Liu M, Freeland J W and Bhattacharya A 2013 Charge transfer and interfacial magnetism in $(LaNiO_3)_n/(LaMnO_3)_2$ superlattices *Phys. Rev.* B **88** 144411

[199] Singh S *et al* 2014 Induced magnetization in $La_{0.7}Sr_{0.3}MnO_3/BiFeO_3$ superlattices *Phys. Rev. Lett.* **113** 047204

[200] Singh S, Xiong J, Chen A P, Fitzsimmons M R and Jia Q X 2015 Field-dependent magnetization of $BiFeO_3$ in an ultrathin $La_{0.7}Sr_{0.3}MnO_3/BiFeO_3$ superlattice *Phys. Rev.* B **92** 224405

[201] Banu N, Singh S, Satpati B, Roy A, Basu S, Chakraborty P, Movva H C P, Lauter V and Dev B N 2017 Evidence of formation of superdense nonmagnetic cobalt *Sci. Rep.* **7** 41856

[202] Banu N, Singh S, Basu S, Roy A, Movva H C P, Lauter V, Satpati B and Dev B N 2018 High density nonmagnetic cobalt in thin films *Nanotechnology* **29** 195703

[203] Speakman J, Rose P, Hunt J A, Cowlam N, Somekh R E and Greer A L 1996 The study of self-diffusion in crystalline and amorphous multilayer samples by neutron reflectometry *J. Magn. Magn. Mater.* **156** 411–2

[204] Gupta M, Gupta A, Stahn J, Horisberger M, Gutberlet T and Allenspach P 2004 Iron self-diffusion in amorphous $FeZr/^{57}FeZr$ multilayers measured by neutron reflectometry *Phys. Rev.* B **70** 184206

[205] Basha M A, Bhatt H, Kumar Y, Prajapat C L, Gupta M, Karki V, Ghosh S K, Basu S and Singh S 2020 Evolution of structural and magnetic properties of FePtCu alloy films on annealing of FePt/Cu multilayers *Phys. Chem. Chem. Phys.* **22** 16107–16

[206] Basha M A, Bhatt H, Kumar Y, Prajapat C L, Gupta M, Basu S and Singh S 2020 Annealing driven interface diffusivity in FePt/Cu multilayer *AIP Conf. Proc.* **2265** 030254

[207] Colgan E G 1990 A review of thin-film aluminide formation *Mater. Sci. Rep.* **5** 1–44

[208] Rothhaar U, Oechsner H, Scheib M and Müller R 2000 Compositional and structural characterization of temperature-induced solid-state reactions in Al/Ni multilayers *Phys. Rev.* B **61** 974–9

[209] Schmidt H, Gupta M and Bruns M 2006 Nitrogen diffusion in amorphous silicon nitride isotope multilayers probed by neutron reflectometry *Phys. Rev. Lett.* **96** 055901

[210] Li M *et al* 2015 Proximity-driven enhanced magnetic order at ferromagnetic-insulator–magnetic-topological-insulator interface *Phys. Rev. Lett.* **115** 087201

[211] Katmis F *et al* 2016 A high-temperature ferromagnetic topological insulating phase by proximity coupling *Nature* **533** 513–6

[212] Swekis P *et al* 2021 Magnetic and electronic properties of Weyl semimetal Co_2MnGa thin films *Nanomaterials* **11** 251

[213] Webster P J 1971 Magnetic and chemical order in Heusler alloys containing cobalt and manganese *J. Phys. Chem. Solids* **32** 1221–31

[214] Dingle R, Störmer H L, Gossard A C and Wiegmann W 1978 Electron mobilities in modulation-doped semiconductor heterojunction superlattices *Appl. Phys. Lett.* **33** 665–7

[215] Moon E J, He Q, Ghosh S, Kirby B J, Pantelides S T, Borisevich A Y and May S J 2017 Structural 'δ doping' to control local magnetization in isovalent oxide heterostructures *Phys. Rev. Lett.* **119** 197204

IOP Publishing

Neutron and X-ray Reflectometry
Emerging phenomena at heterostructure interfaces
Saibal Basu and Surendra Singh

Chapter 4

Correlation of interface morphology and magnetism in heterostructures: off-specular (diffuse) scattering

4.1 Off-specular (diffuse) scattering

Neutron scattering plays an important role in studying the nanomagnetic systems either consisting of artificially arranged multilayers or laterally patterned islands, or a combination of both. As discussed in chapter 3, in the case of specular reflectivity (x-ray reflectivity (XRR) and neutron reflectivity (NR)), the scattering vector Q is perpendicular to the film surface and thus can only provide the depth profile of the structure in thin films or multilayers. The depth profiles of the structure and magnetic properties of the thin film, obtained from specular reflectivities, are averaged out in the plane of the film. However, interfaces have lateral fluctuations in the SLD in the form of magnetic domains and lateral height fluctuation in the form of roughness between layers. In such cases, it is interesting to look at the scattered intensities around the specular reflection direction (angle of incidence $\theta_i \neq$ angle of reflection θ_r) (figure 4.1) and probe the in-plane structure and magnetism. The intensity pattern measured for studying the in-plane information at interfaces is called off-specular (diffuse) scattering (figure 4.1). In the off-specular geometry (figure 4.1), a small in-plane component (Q_X) of the resultant scattering wave vector Q appears in addition to the out-of-plane component (Q_Z). It is noted that at grazing incidence of the probe (neutron/x-ray) this in-plane component, Q_X of the scattering wave vector is very small (of the order of 0.1–50 μm^{-1}) and therefore it will be sensitive to a characteristic size ranging from 0.1 to 50 μm of in-plane lateral structure in the real space. These sizes correspond to typical sizes of micro-magnetic domain structures and thus magnetic off-specular scattering is mostly used to probe such problems.

Experimentally, off-specular measurements are usually performed by using a linear one-dimensional (1D) or 2D position-sensitive detector (PSD) set after the sample and by measuring the scattering on the detector as a function of the incidence angle. Using 2D PSD a 2D mapping in the reciprocal space (a $Q_X - Q_Z$ intensity

doi:10.1088/978-0-7503-4695-5ch4

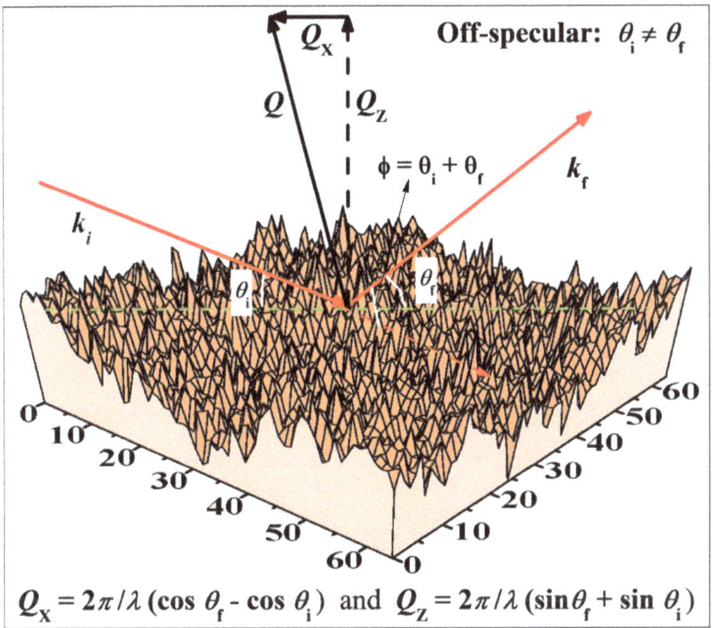

Figure 4.1. Off-specular scattering from a rough surface and related geometry. The scattering vector Q is not perpendicular to the thin-film plane and it is the summation of the perpendicular (out of the plane of the surface) component (Q_Z) and parallel (in-plane) component (Q_X).

map) can thus be obtained and intensity scattered along the Q_Z (specular reflectivity) and Q_X directions contains information about the depth-dependent structure and in-plane micrometric structures, respectively.

Off-specular or diffuse x-ray scattering facilities are easily accessible due to the availability of high-intensity x-ray laboratory sources and very bright synchrotron sources. Off-(diffuse)neutron scattering has also contributed significantly during the last two decades because of the availability of high brilliance neutron sources, advancements in beamline optics, and gains in detector efficiency combined with advances in the theoretical description and analysis of off-specular scattering technique.

In an experiment, the specular reflectivity $[\theta_i = \theta_f = \theta]$ is measured as a function of the wave vector transfer component Q_Z, given as $Q_z = \frac{2\pi}{\lambda}[\sin(\theta_i) + \sin(\theta_f)] = \frac{4\pi}{\lambda}\sin\theta$, with λ as the wavelength of the probe (x-ray/neutron) [1–5]. The in-plane wave vector transfer $Q_x = \frac{2\pi}{\lambda}\left[\cos(\theta_f) - \cos(\theta_i)\right]$ contributes to the off-specular (diffuse) scattering $(\theta_i \neq \theta_f)$ for determining the in-plane structure and morphology [2, 3, 5]. The scattering geometry in reciprocal space for different scans [4] of reflectivity is shown in figure 4.2. The reflectivity along Q_Z with $\theta_i = \theta_f$ and $Q_X = 0$ denotes the specular reflectivity. As discussed in chapter 3, the depth-dependent magnetic structure is obtained from specular ($Q_X = 0$) PNR with polarization analysis, i.e. non-spin-flip (NSF), R^{++} and R^{--}, and spin-flip (SF), R^{+-} and R^{-+}, reflectivities, in the form of the magnitude (M) and direction (α) of the magnetization vector with respect to the polarization of neutron in the multilayer [2, 3, 6] perpendicular to Q_Z. NSF probes the projection of the

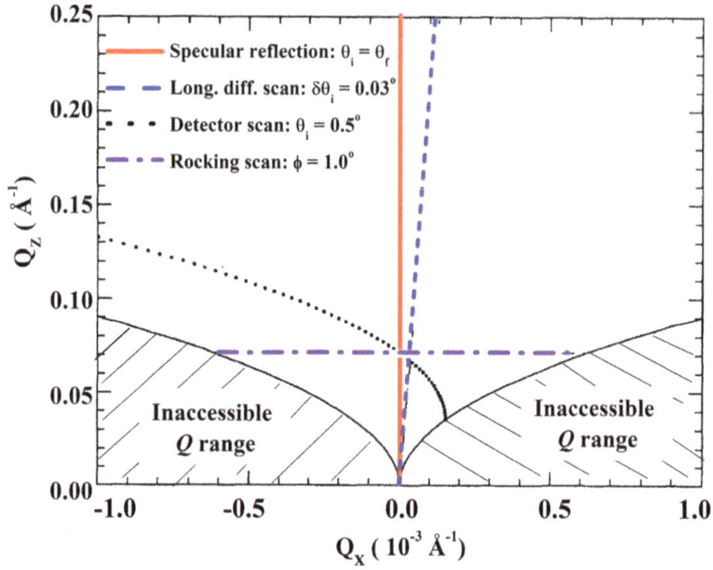

Figure 4.2. Different scans in reciprocal space (Q_X–Q_Z map). The solid (red) line represents a specular reflectivity ($\theta_i = \theta_f$, $Q_X = 0$) and the inclined dashed (blue) line is a longitudinal diffuse scan with an offset $\delta\theta_i = 0.03°$. The dash–dotted (violet) line is a rocking scan with a scattering angle $\phi = \theta_i + \theta_f = 1°$. The detector scan with incidence angle $\theta_i = 0.5°$ is given by the dotted (black) line. Reproduced with permission from [4]. Copyright 1995 the American Physical Society.

magnetization vector parallel ($M_{||}$) to the in-plane polarization direction, while SF is sensitive to the in-plane perpendicular component ($M_\perp$), both of which are perpendicular to Q_Z. There are three different types of off-specular scattering, which are a projection of intensity along different reciprocal spaces ($Q_X - Q_Z$ map) and termed as (i) longitudinal diffuse scan, (ii) rocking diffuse scan, and (iii) detector diffuse scan. A longitudinal diffuse scan is a nearly specular scan where the angle of reflection (θ_f) is slightly out of the specular condition with $\theta_f = \theta_i - 2\delta\theta_i$ and in reciprocal space this scan lies on a straight line that is inclined at an angle $\delta\theta_i$ against the Q_Z direction (blue dashed line in figure 4.2). A rocking scan is performed by rotating the sample at a fixed detector position. Thus, the angle of incidence θ_i varies and the scattering angle $\phi = \theta_i + \theta_f$ is constant and these scans are nearly Q_X scans at a fixed Q_Z position (violet, dash–dotted horizontal line in figure 4.2). A detector scan is performed at a fixed angle of incidence (θ_i) and a varying scattering angle ϕ, and both the Q_X and the Q_Z components of the scattering vector are changed simultaneously. The path of this scan in reciprocal space is a parabola (a dotted line, black, in figure 4.2).

The shaded regions (inaccessible Q range) in reciprocal space shown in figure 4.2 are not accessible with the experimental set-up due to geometrical restrictions. Reflectivity measurements from a thin film both in specular and off-specular mode characterize the sample for structural and morphological properties. Off-specular x-ray scattering provides the height–height correlation at different interfaces and for a self-affine fractal surface it is defined as [1] $C(x, y) = \langle \delta z(0)\delta z(x, y)\rangle = \sigma^2 \exp\left(-\left[\frac{r}{\xi}\right]^{2h}\right)$, where σ, h, and ξ are a root mean square roughness (a true height–height fluctuation), the Hurst parameter, and the in-plane height–height correlation length, respectively. The exponent $0 < h < 1$ determines the fractal dimension ($D = 3 - h$) of the interface (i.e. how jagged the interface is; the larger the value of h the smoother is the interface). To fit off-specular XRR data, the morphological parameters (σ, h, and ξ) at the interface were varied while keeping the thickness and ESLD parameters obtained from specular XRR (NR) constant.

Figure 4.3 shows the simulated specular XRR and diffuse x-ray scattering in different modes for a Fe film of thickness ~40 nm grown on a Si substrate. Figure 4.3(a) shows the specular XRR and the longitudinal off-specular x-ray scattering assuming an offset angle ($\delta\theta_i$) of 0.1°. The inset of figure 4.3(a) shows the electron scattering length density depth profile of the thin film. Different parameters used for simulating reflectivity profiles are given in table 4.1. Figures 4.3(b) and (c) show the diffuse XRR for detector and rocking scans, respectively, at different angles of incidence in specular XRR. Thus, x-ray reflectivity in specular and off-specular mode provides detailed structure and morphology of different interfaces. This is equally true for diffuse polarized neutron scattering (DPNS). Moreover polarized DPNS reveals magnetic domain correlations in a magnetic film. Unlike atomic force microscopy, which provide morphology of the surface at different length scales, diffuse x-ray scattering provides morphology of different interfaces in

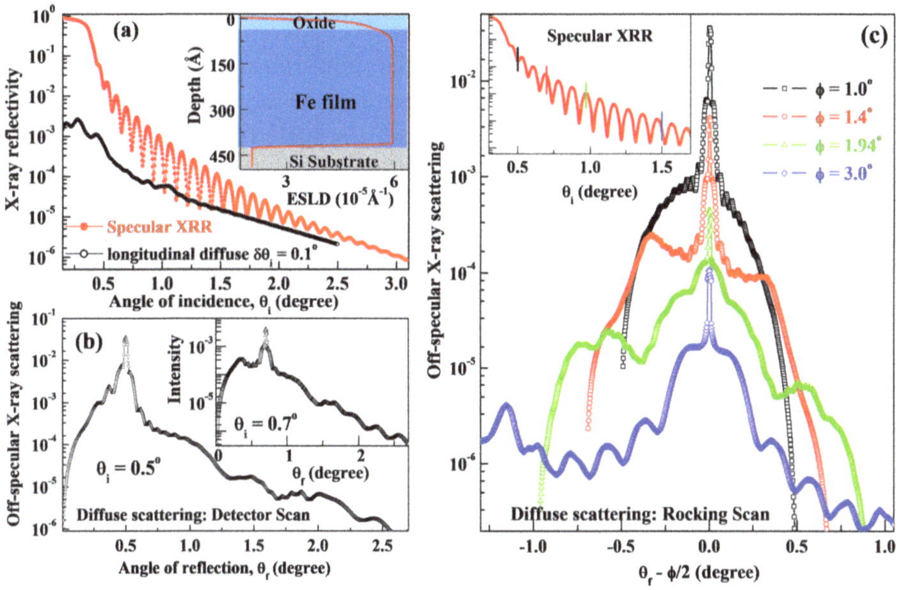

Figure 4.3. (a) Simulated specular XRR and longitudinal off-specular XRR profile as a function of angle of incidence for a Fe film (thickness ~40 nm) on Si substrate. The inset shows the electron scattering length density (ESLD) depth profile of the film. (b) Off-specular x-ray scattering profile in detector scan at two angles of incidence, e.g. $\theta_i = 0.5°$ and 0.7°. (c) Simulation of off-specular x-ray scattering profile in rocking scan for different values of scattering angle ϕ (= $\theta_i + \theta_f$). The inset shows the specular XRR and the vertical lines show the position of the angle at which different rocking diffuse scans were simulated.

Table 4.1. Different parameters of the Fe film used for simulation of specular XRR and diffuse x-ray scattering.

Layer	Thickness (Å)	ESLD (10^5 Å^{-2})	Roughness, σ (Å)	Hurst parameter, h	Correlation length, ξ (Å)
Fe oxide	20	5.05	7	0.2	600
Fe	400	5.97	25	0.2	10 000
Si substrate	—	2.00	3	0.5	40 000

heterostructures and play an important role to study the structure–property correlation in these systems.

4.2 Off-specular (diffuse) x-ray scattering

We present a description for off-specular x-ray scattering since this technique has been used heavily to characterize interface morphology in many heterostructures. It will be followed by a discussion on off-specular (diffuse) neutron scattering for studies in physical and magnetic interface morphology. For both x-ray and neutron off-specular scattering, x-ray diffraction (XRD) is often a prerequisite and is done routinely. The theoretical analysis of off-specular scattering is given in chapter 2. Off-specular x-ray scattering has been widely used to extract in-plane morphology (height–height correlation) at the interfaces [1]. Investigation of the surface phenomena in liquids has been one of the important surface studies using synchrotron-based x-ray diffuse scattering during the last three decades. Sinha *et al* [1] have demonstrated theoretically that diffuse scattering from liquid surfaces diverges algebraically at the specular reflection angle, creating large tails extending into the surrounding off-specular region. This theoretical prediction was later verified in several experimental studies on simple molecular liquids [7, 8] and liquid metals [9, 10]. Diffuse x-ray scattering was also used to study the morphological properties of different interfaces in polymers [11], biological membranes [12], Langmuir–Blodgett (LB) films [13–15], organic semiconductors [16], insulators [17], semiconductors [4], and magnetic multilayers and heterostructures [18, 19].

Temperature-dependent x-ray scattering (specular, off-specular reflectivity and XRD) measurements from a single-crystal $SrTiO_3$ surface suggested the formation of surface rumples (a wavy surface) over micron+ length scales below the structural phase transition at 105 K, which is strongly correlated with the atomic structure [17]. The evolution of interface morphology on annealing a Ni/Ti multilayer, which affects the kinetics of alloy formation at interfaces, has been studied using diffuse XRR [18], and a combined study of specular and off-specular reflectivity suggested a modification in interface morphology as well as an increase in in-plane correlation length on increasing the annealing temperature.

In addition, hard (non-resonant) x-ray (energy ~5 keV) scattering is mostly used to study in-plane morphology in the form of height–height correlation and it cannot be used to study the magnetic properties because non-resonant magnetic x-ray scattering amplitudes are about three orders of magnitude smaller than those of pure charge x-ray scattering [20]. However, if the x-ray energy is tuned to a transition involving magnetically active levels of a particular atom (mostly in a soft x-ray energy range) in the sample, the charge and magnetic scattering amplitudes become comparable [20]. Thus using a soft x-ray near the resonance edge of the atom an x-ray resonant magnetic scattering (XRMS) in specular and diffuse scattering from the surface can be employed to determine both the structural (i.e. charge) and magnetic morphology (roughnesses, correlation length) of magnetic thin films. Experimentally, the scattered intensity for incident helicity parallel (I^+) and antiparallel (I^-) to the sample magnetization (~an applied external field) is measured and the average of the intensity $[(I^+ + I^-)/2.0]$ is taken to determine the charge (structural) roughness and the difference ($I^+ - I^-$) is used to determine the magnetic roughness [21, 22]. Lee *et al* [23] formulated a theoretical analysis of XRMS from rough surfaces and interfaces for diffuse (off-specular) scattering and general expressions were derived under Born approximation (BA) and under distorted-wave Born approximation (DWBA) [1] for both single and multiple interfaces. Thus using the specular and diffuse soft x-ray scattering technique at the resonance edge of the element one can study the element-specific structure and magnetic morphology at interfaces in the thin-film heterostructures.

4.2.1 Interface morphology of soft matter heterostructures

The interplay of ordering, confinement, and growth in ultrathin soft matter films gives rise to various interesting phenomena and such systems also enhance our understanding of the fundamental interactions of amphiphilic molecules important for biological systems. Ordered metal–organic films, deposited using the Langmuir–Blodgett (LB) technique, are one of the ideal systems which provide an opportunity to study chemical reactions and phase transitions in confined geometries. The advent of grazing incidence x-ray scattering techniques has enabled us to probe the interfacial structure of these soft matter heterostructures at a very high resolution and as a result has improved our knowledge about the mechanism of growth processes and physical–chemical–morphological properties of these ultrathin heterostructures. Specular XRR and diffuse scattering studies from solid–solid interfaces have been described very well assuming self-affine fractal [1] surfaces. X-ray scattering studies from LB films have shown that the interfacial correlation can vary from self-affine fractal [14, 15] nature, observed in diverse physical systems [1], to long-range logarithmic [24], characteristic of capillary waves on liquid surfaces [8]. Although off-specular x-ray scattering is an excellent tool for interface morphology, due to low contrast for the soft matter and biological systems and limited intensity for laboratory-based x-ray sources, x-ray off-specular scattering measurements from these systems have been used in a limited way. However high-intensity synchrotron x-ray sources have enabled better counting statistics over a wider range in momentum space providing a better spatial resolution.

4.2.2 Vertically correlated interface morphology in an organic semiconductor multilayer

Organic semiconductors are solids with building blocks of π-bonded molecules or polymers made up of carbon and hydrogen atoms as well as heteroatoms such as nitrogen, sulphur, and oxygen, at times. The family of phthalocyanines (Pc) are important molecules in the field of molecular electronics because of their thermal stability and the possibility to tune their structure, chemical, magnetic, and transport properties by employing different metallic cations within the Pc molecular cage. The transition metal phthalocyanines (TMPcs) are archetypal organic semiconductors with promising applications such as dyes, field-effect transistors, gas sensors, field emission applications, light-emitting diodes, solar cells, and single-molecule-devices [25–27]. TMPcs are planar molecules with a transition metal ion core surrounded by organic rings (see the inset of figure 4.4(b)) and they provide an excellent way to replace the metal ion and modify the metal ion's spacing using different substrates and preparation methods to modify the electronic and magnetic properties of the TMPc. Considerable interest in the magnetic properties of TMPc films, e.g. manganese phthalocyanine (MnPc), iron phthalocyanine (FePc), and cobalt phthalocyanine (CoPc), has been present in the last two decades. Gredig *et al* [28] observed ferromagnetism in a FePc film below 4.5 K due to inter-iron–chain interactions and

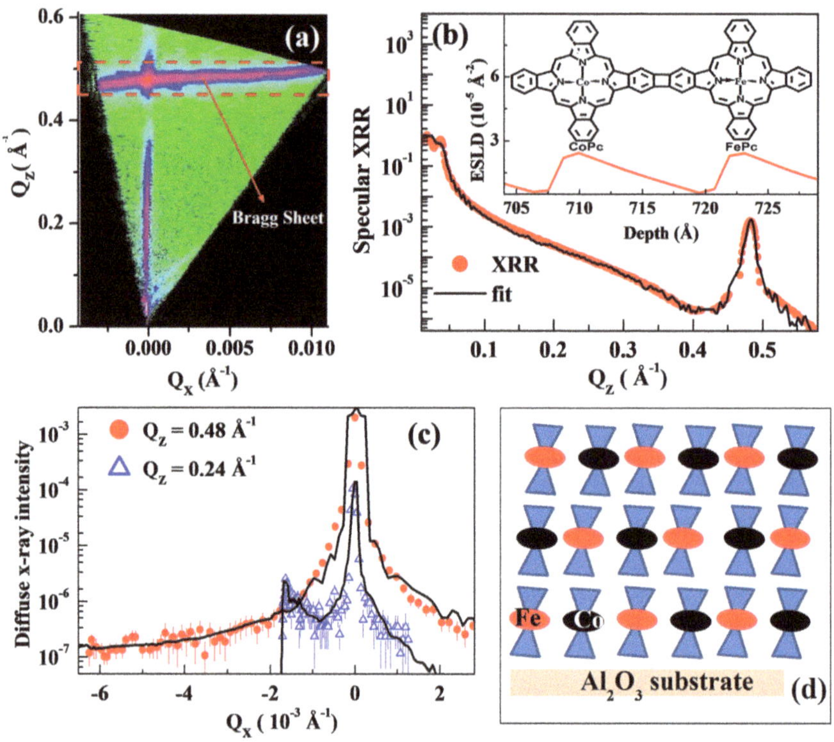

Figure 4.4. (a) Reciprocal space map (Q_X–Q_Z intensity map) of x-ray reflectivity for (Co–Fe)Pc film. (b) Specular x-ray reflectivity (XRR) measurements for binuclear (Co–Fe)Pc film. The inset shows the electron scattering length density (ESLD) depth profile across the dimer of binuclear (Co–Fe)Pc film which gave the best fit to the XRR data. The inset also shows the representation of the binuclear (Co–Fe)Pc molecule, which makes a dimer of CoPc and FePc. (c) Off-specular (diffuse) x-ray scattering (detector scan) from binuclear (Co–Fe)Pc film at two Q_Z values. (d) Representation of (Co–Fe)Pc film grown on a sapphire substrate. Reproduced with permission from [16]. Copyright 2014 American Chemical Society.

found that the interactions were strongly influenced by the growth of the film on different substrates. Similarly, different magnetic properties, e.g. loss and quenching of the magnetic moment of Co ions for CoPc molecules, are also reported. The TMPc films are expected to bridge two novel disciplines of spintronics and molecular electronics, because of their unique electronic and magnetic characteristics arising from the transition metal atoms in their framework. Kumar *et al* [29] observed very high mobility of 110 $cm^2 V^{-1} s^{-1}$ at room temperature for a binuclear (Co–Fe)-phthalocyanine [(Co–Fe)Pc] films of thickness ~200 Å, which was attributed to the structural ordering of (Co–Fe)Pc dimers. In the case of a (Co–Fe)Pc dimer (inset of figure 4.4(b)), CoPc and FePc molecules are covalently joined through the corner benzene ring via two C–C bond formations.

Using XRR and polarized neutron reflectivity (PNR), Singh *et al* [16] studied the depth-dependent chemical composition and magnetization of pure FePc, CoPc, and binuclear (Co–Fe)Pc thin films grown on sapphire substrates. The specular reflectivity results suggested that the structural changes in binuclear (Co–Fe)Pc film compared to the pure film are responsible for a huge reduction in resistivity and the presence of ferromagnetism in this film at low temperatures. Specular PNR data revealed that the binuclear (Co–Fe)Pc film is ferromagnetic with a magnetization of 50 ± 15 kA m^{-1} at 10 K, but the pure FePc or CoPc films showed negligible magnetization at 10 K. Diffuse x-ray scattering provided detail morphology of the layer structure suggesting highly correlated interfaces along the normal and in the plane of the film. Figure 4.4(a) shows the diffuse x-ray scattering (Q_X–Q_Z intensity map) for the binuclear (Co–Fe)Pc film [16]. The intensity along Q_Z at $Q_X = 0.0$ Å is the specular XRR and is shown in figure 4.4(b) along with the fit (solid line) to XRR. The x-ray intensity as a function of Q_X at fixed Q_Z (horizontal line profile in figure 4.4(a)) is off-specular (diffuse) x-ray scattering (detector scan).

Specular XRR provided the total thickness of the film ~785 ± 8 Å and also exhibited a Bragg peak at Q_Z of ~0.49 Å^{-1}, which corresponds to a bilayer thickness (~$2\pi/Q_Z$) of ~12.9 Å for binuclear (Co–Fe)Pc films. The inset of figure 4.4(b) shows the electron scattering length density (ESLD) depth profile (solid red line) across the (Co–Fe) dimer and represents the organic (small SLD) and metallic (large SLD) parts of the molecule. The dimer depicts two periods of a bilayer structure (inset of figure 4.4(b)). Thus, specular XRR suggested that within a film the molecules are self-assembled in a multilayer (repetition of bilayer along the thickness) structure with a repetition of a bilayer (larger lattice spacing ≈ 14 Å) structure, which is aligned normal to the surface. The bilayer indicates that the film has grown such that the shorter (lattice spacing ≈ 3.5 Å) molecular *b*-axis is oriented parallel to the substrate plane, i.e. edge-on stacking, as shown schematically in figure 4.4(d) [16]. XRD measurements also confirmed the same [16].

In addition, the (Co–Fe)Pc molecules are highly correlated along the out-of-plane direction which is evident from the presence of high x-ray scattering intensity (the Bragg sheet) as a function of Q_X at fixed Q_Z (corresponds to the Bragg peak) as shown in figure 4.4(a) (highlighted in dashed rectangle). Formation of the Bragg sheet suggests a high correlation of molecules (bilayer) along the thickness of the film over a length scale of the total thickness of the film (~1000 Å). Figure 4.4(c) shows the diffuse x-ray scattering at two values of Q_Z. To study the correlation of atoms in the plane of the film, the diffuse XRR data were analyzed using a self-affine fractal surface model [1, 2, 16]. Solid lines in figure 4.4(c) show the fit to diffuse data suggesting an in-plane correlation length (ξ) of 0.30 ± 0.05 μm, for the interfaces of binuclear (Co–Fe)Pc film.

4.2.3 Correlation between surface rumpling, structural phase transformation, and surface morphology in $SrTiO_3$

Strontium titanate ($SrTiO_3$) is a perovskite-structured oxide with remarkable multi-functional properties and in stoichiometric form it is an optically transparent insulator with a high dielectric constant. $SrTiO_3$ (STO) has many attractive features as a substrate for growing epitaxial complex oxide thin film, e.g. low cost compared to other perovskite single crystals, the relative absence of twins, a lattice parameter that is reasonably well-matched to many perovskite films, and a nonpolar structure that is conducive to smooth surfaces. STO undergoes a structural phase transition from cubic ($a = 3.905$ Å) to tetragonal ($c/a = 1.000\,56$) phase at ~105 K [30]. The phase transition involves the rotation of TiO_6 octahedra and has been featured as a classic example of a soft-mode phase transition. However, a disadvantage of STO is that the cubic-to-tetragonal structural phase transformation may induce strain or microcracks in a film grown on an STO substrate, which may influence strain-sensitive properties, e.g. magnetoresistance, multiferroicity, ferroelectricity, etc. Using temperature-dependent x-ray reflectivity (specular and diffuse) and atomic force microscopy, Singh *et al* [17] studied the surface structure of STO and found severe rumpling of the STO surface in addition to the change in atomic structure at 105 K. The rumpling transformation is correlated with the cubic-to-tetragonal phase transformation and the rumpling transformation is reversible. The observations were also correlated with those of the bulk structure inferred from XRD. Thus the preponderance of surface rumpling of STO may be a source of concern for studies of strain-sensitive properties of the films grown on STO substrate [17]. The report on the STO surface also suggested that the surface rumpling is accompanied by a change in the bond hybridization of Ti–O at the surface and thus may modify the interaction at the interface with the oxide film grown on the STO surface.

Figures 4.5(a) and (b) show diffuse XRR (Q_X–Q_Z scattering intensity map at a small angle) from the (001) surface of the STO single crystal at temperatures of 300 and 105 K, respectively [17]. A sharp specular reflection above 150 K was observed for the sample and a small broadening of the specular reflection about $Q_X = 0$ was first observed at a temperature of 150 K, and with further cooling the specular reflection broadened dramatically [17]. The x-ray intensity as a function of Q_Z at $Q_X = 0$ corresponds to the specular XRR. The specular XRR data for an STO single crystal at 300 and 105 K is shown in figure 4.5(c) and the XRR intensity at 105 K is reduced at larger Q_Z at 105 K, suggesting an increase in surface roughness at lower temperature. The x-ray intensity as a function of Q_X at fixed Q_Z corresponds to diffuse x-ray scattering. It is evident from figures 4.5(a) and (b) that, while cooling the sample, the specular reflection (centred about $Q_X = 0$) broadens along Q_X and

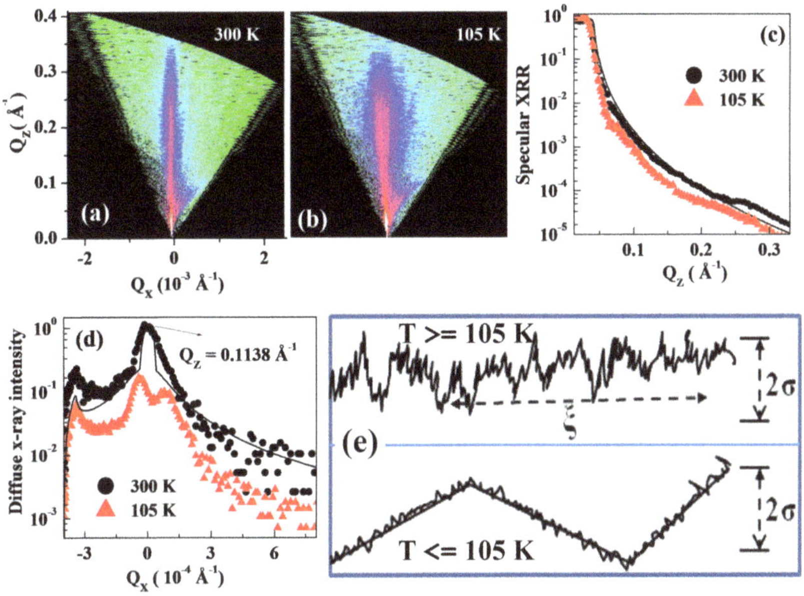

Figure 4.5. Reciprocal space map (Q_X–Q_Z intensity map) of x-ray reflectivity for $SrTiO_3$ (STO) single crystal surface at (a) 300 K and (b) 105 K. (c) Specular x-ray reflectivity (XRR) measurements for STO at different temperatures. The solid line is fit to XRR data at 300 K. (d) Off-specular (diffuse) x-ray scattering (detector scan) from STO at different temperatures at a Q_Z of 0.1138 Å^{-1}. Diffuse x-ray scattering data at different temperatures in (d) are shifted vertically for clear visibility. (e) Representation of surface modification for STO surface inferred from x-ray scattering measurements above (upper panel) and below (lower panel) 105 K.

splits into multiple reflections indicating significantly increased in-plane inhomogeneity. The diffuse x-ray scattering data for the sample at 300 and 105 K, as a function of Q_X for $Q_Z \sim 0.1138$ Å^{-1} is plotted in figure 4.5(d). It is evident from figure 4.5(d) that the behaviour of the diffuse x-ray intensity for large Q_X is affected very little by temperature. Using a self-affine fractal surface model [1], the diffuse scattering data of the (001) surface of STO for large Q_X was fitted (solid line in figure 4.5(d)) for the corresponding morphological parameters of σ (root mean square, rms, roughness) = 4 ± 1 Å, h (Hurst parameter) = 0.45 ± 0.05, and ξ (in-plane correlation length) = 0.8 ± 0.05 μm. The corresponding surface morphology for the STO at higher temperatures ($T > 105$ K) is depicted in figure 4.5(e) (upper panel). The broadening and splitting of the reflected x-ray intensity at low temperatures (mostly below 105 K) in the Q_X–Q_Z scattering maps were observed which was the consequence of the incoherent superposition of reflection from surfaces with long waviness (surface facets). Facets tilted with respect to each other as depicted in the lower panel of figure 4.5(e) are an example of a rumpled surface that could produce the splitting of the specular reflection shown in figures 4.5(b) and (d). The angular spread of the facets is not more than 0.5°.

The x-ray scattering data also suggested that the surface transition from smooth to the rumpled surface was reversible with temperature, i.e. the reflection sharpened at high temperatures [17]. Detailed temperature-dependent x-ray scattering study on the STO surface suggested that the rumpling of the surface is correlated with a change of atomic structure, as observed by a degradation of the mosaic quality of the crystal substrate (inferred from the rocking curve width), and increased the microstrain and/or decreased the grain size of the crystal (inferred from the radial width of a Bragg reflection).

4.2.4 Correlation of charge (structure) and magnetic roughness in Fe/Gd multilayers

Osgood *et al* [22] described a general method for calculating XRMS in both specular and diffuse mode in the BA, by treating thin-film systems as having separate, but correlated, rough, magnetic, and structural surfaces. Under BA they suggested that the diffuse scattering intensity differs between left- and right-circularly polarized x-rays for a rough surface and it vanishes unless the structural and magnetic roughnesses are correlated. Nelson *et al* [21] extended the BA approach suggested in [22] for studying charge–magnetic correlation in the Fe/Gd multilayer. The photon energy was tuned to 7245 eV (Gd L_3 edge) for this experiment.

The structural parameters (thickness, interface roughness, and ESLD) of the multilayer were estimated by measuring the specular reflectivity (average of the intensity measured with opposite photon helicities, i.e. $[(I^{+}+I^{-})/2.0]$) with circularly polarized x-rays. Figure 4.6(a) shows the specular reflectivity for the Fe/Gd multilayer with a structure of Si (substrate)/[Gd (52 Å)/Fe (35 Å)]$_{15}$/Si (20 Å), where 15 is the number of Ge/Fe bilayers. The peaks at different Q_Z corresponding to the multilayer Bragg peaks. Figures 4.6(b) and (c) show the diffuse sum $[(I^{+}+I^{-})/2.0]$ and difference $(I^{+}-I^{-})$ data in the rocking scan for two angle of incidences corresponding to a Q_Z of ~0.147 and 0.217 Å^{-1}, which are the position of second and third order Bragg peaks, respectively, as shown in figure 4.6(a). The solid lines in figures 4.6(b) and (c) represent the best fit to data suggesting a charge–charge (ξ_{cc}) and charge–magnetic (ξ_{cm}) correlation length of 230 and 1420 Å, respectively [21].

Within the DWBA approach, Lee *et al* [23] extended the theoretical analysis under BA provided in [21, 22], and generalized the conventional theory of ordinary (Thomson) x-ray scattering from surfaces possessing self-affine structural roughness to the case of resonant magnetic x-ray scattering from surfaces or interfaces of ferromagnetic materials possessing both structural and magnetic roughnesses. Under DWBA, an improved fit to diffuse XRMS data for the Fe/Gd multilayer, shown in figure 4.6, was obtained with almost similar morphological parameters [23].

4.2.5 Correlation between charge and magnetic morphology in manganite film

Another important study, where x-ray scattering in both the hard and soft energy windows was applied to measure specular and diffuse scattering data from a complex oxide manganite thin-film system, was reported in [31, 32]. Combining electrical transport (resistance) measurements with different x-ray scattering measurements such as x-ray absorption spectroscopy (XAS), x-ray magnetic circular dichroism (XMCD), specular/off-specular XRMS, and specular and diffuse hard (non-resonant) x-ray scattering measurements as a function of temperature, Singh *et al* [31, 32] elucidated relationships between the electronic, magnetic, and

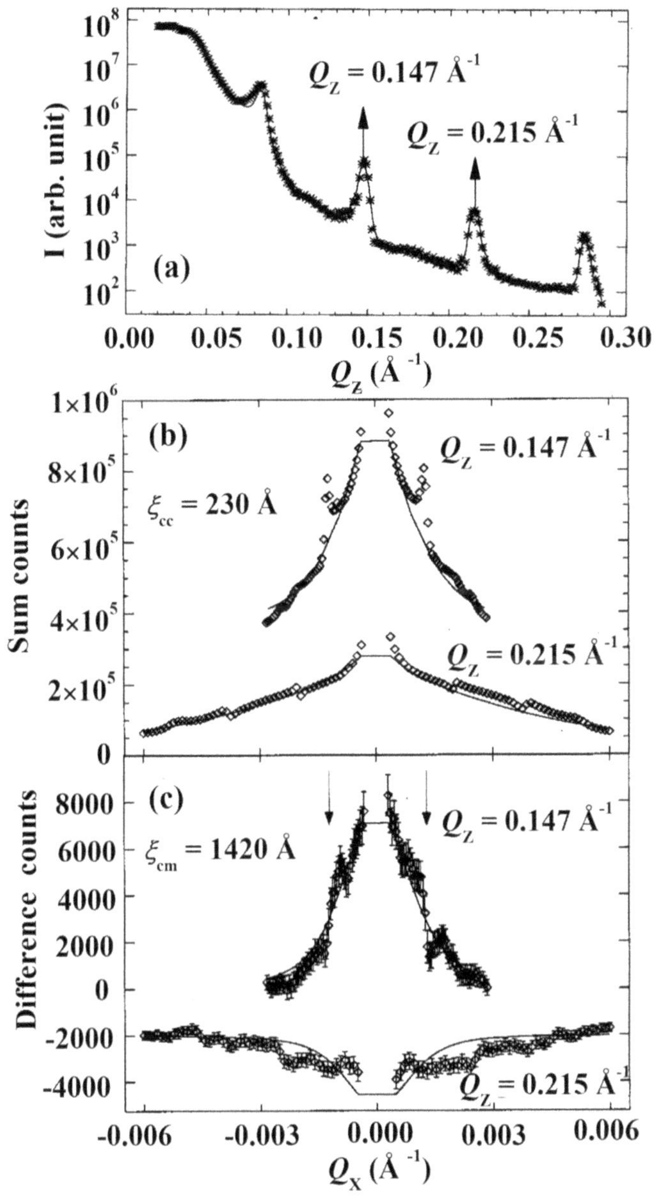

Figure 4.6. (a) Specular x-ray reflectivity for Fe/Ge multilayer. Diffuse XRMS rocking curve data with the (b) sum $[(I^{+}+I^{-})/2.0]$ and (c) difference $(I^{+}-I^{-})$ of opposite photon helicity from the multilayer. The photon energy was tuned at 7245 eV (Gd L_3 edge). The sum and difference data provide the charge–charge (structure) and charge–magnetic roughness correlation at interfaces. The data in (b) and (c) are shifted vertically for better visualization. Reproduced with permission from [21]. Copyright 1999 the American Physical Society.

morphological structure of the thin films of $(La_{0.4}Pr_{0.6})_{1-x}Ca_xMnO_3$ (LPCMO) manganite with $x = 0.33$ (film S1) and 0.375 (film S2). These LPCMO systems in the bulk phase have shown drastically different magnetic and transport properties due to a change in the internal strain because of a small change in doping concentration. The thin films showed different metal-to-insulator transition (MIT) temperatures (T_{MI} = 51 and 109 K, for LPCMO with x = 0.33 and 0.375, respectively) and magnetic transition temperatures (T_C ~ 130 and 150 K for LPCMO with x = 0.33 and 0.375, respectively) [32]. Using off-specular XRMS, different charge–charge (electronic) and charge–magnetic (magnetic) correlation lengths were estimated for these LPCMO films across the MIT temperature [32]. The different charge–magnetic correlation length (~domain size) for the two films increases below the MIT and the different correlation lengths shown by the two films may be responsible for different macroscopic (transport and magnetic) properties. The combination of hard x-ray (Cu K_{α}) and soft XRMS diffuse scattering also suggested a constant charge–charge correlation length across the whole temperature range.

Figures 4.7(a) and (b) show the XAS and XMCD measurements from LPCMO films S1 and S2, respectively, at 20 K and near the Mn L edge in an applied magnetic field of 700 Oe [32]. Both films are grown of similar thickness on NGO substrate using pulsed laser deposition. The maximum of the (negative) XMCD signal at the Mn L_3 edge can be seen at ~640.5 eV, which was the energy used for the off-specular XRMS measurements for sum and difference intensity shown in figures 4.7(c)–(f). It has been shown that the diffuse difference ($I^{+}-I^{-}$) intensity (charge–magnetic scattering) results predominantly from charge–magnetic cross-correlations while the diffuse sum ($I^{+}+I^{-}$) intensity (charge scattering) results predominantly from charge–charge correlations [21, 23]. Figures 4.7(c) and (d) show the experimentally measured diffuse charge scattering data and corresponding fit as a function of Q_X, at fixed Q_Z = 0.113 Å^{-1}, at an energy of 640.7 eV, from S1 and S2, respectively. The diffuse charge scattering data at different temperatures across MIT remain constant suggesting that the charge–charge correlation length does not vary with temperature [32]. The inset shows the hard (non-resonant) specular XRR from the two films. A comparison of diffuse charge–magnetic scattering (difference intensity) data (symbol) and corresponding fits (solid lines) at 70 K and 50 K for S1 and S2 are shown in figures 4.7(e) and (f), respectively. The in-plane charge–charge (structure) and charge–magnetic correlation length as a function of temperature obtained from temperature-dependent diffuse XRMS data from two films are shown in figures 4.7(g) and (h).

4.3 Off-specular (diffuse) neutron scattering

While specular PNR provides the detailed depth-dependent magnetic structure in the magnetic heterostructure, off-specular (diffuse) polarized neutron scattering has played an important role in studying in-plan fluctuation of magnetic roughness (morphology) and its correlation, including (i) diffuse scattering from antiferromagnetic Fe/Cr superlattices [33]; (ii) in-plane magnetic disorder and its correlation in Co/Cu multilayer [34]; (iii) quantifying domain structure in magnetic layers [35]; (iv) magnetic properties of a large array of in-plane magnetized FM Co disks (where peaks in off-specular intensities were observed due to the lateral periodicity of the disk array) [36]; (v) nonuniform and highly correlated twisted canted state in Fe/Cr multilayer in the spin-flop phase [37]; (vi) magnetic fluctuations of magnetization

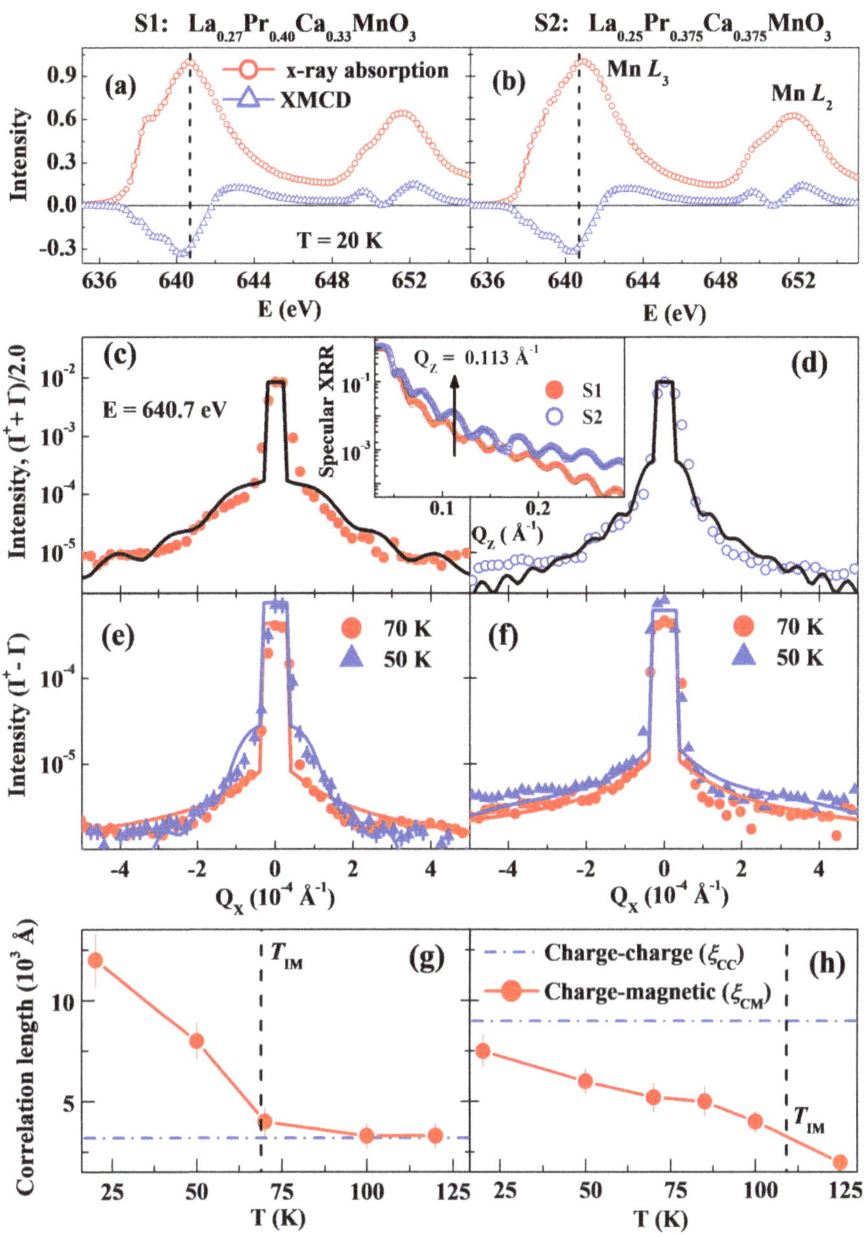

Figure 4.7. Near-surface x-ray absorption, XAS $[(I^+ + I^-)/2.0]$ and XMCD $(I^+ - I^-)$ spectra for LPCMO thin films (a) S1 and (b) S2 at 20 K near the Mn L edge (~640 eV). Diffuse charge scattering data, $(I^+ + I^-)/2$, as a function of Q_X from (c) S1 and (d) S2 at a resonant energy of 640.7 eV ($\sim Q_Z = 0.113$ Å^{-1}). The inset shows the specular hard x-ray reflectivity (Cu K_α radiation) pattern from two films and a vertical line indicates the value of Q_Z at which we have also carried out diffuse hard x-ray scattering measurements. Diffuse charge–magnetic scattering measurements $(I^+ - I^-)$ and corresponding fits (solid lines) at 70 and 50 K (below MIT) as a function of Q_X at fixed $Q_Z = 0.113$ Å^{-1} (soft x-ray energy = 640.7 eV) from (e) S1 and (f) S2 lPCMO films. In-plane charge–charge (ξ_{CC}) and charge–magnetic (ξ_{CM}) correlation length as a function of temperature across the insulator-to-metal transition (T_{IM}) obtained from diffuse x-ray scattering data are shown in (g) S1 and (h) S2 lPCMO films. Reproduced with permission from [32]. Copyright 2016 Macmillan.

reversal in exchange-biased multilayers [38–40]; and many more. Diffuse polarized neutron scattering has also been used to study the reversal process in neutron polarizing super-mirrors [41], waveguide structures [42] and in patterned or self-organized magnetic micro-structures [43–45].

Using unpolarized diffuse neutron scattering in addition to specular neutron reflectivity, Singh *et al* [46, 47] studied the detailed corrosion behaviour of Ni films. The study also related the morphological parameters of a Ni surface obtained from neutron scattering and atomic force microscopy and found a close match. Later diffuse polarized neutron scattering was used to study the separation and correlation of structural and magnetic roughness in a Ni thin film [48]. Paul *et al* [49] probed the vertically correlated magnetic structures in Er/Tb multilayer using off-specular polarized neutron scattering and found the existence of magnetic vortex-like domains associated with magnetic helical ordering within the Er layers. Recently, diffuse polarized neutron scattering with polarization analysis was successfully applied to investigate the interface morphology and its correlation in Gd/Co multilayers [50–52]. Therefore the use of neutron reflectivity both in specular and off-specular geometry provides complete structure and morphology of the magnetic system and is very important for understanding the new magnetic phases and behaviours at different interfaces in a heterostructures. Here we will describe a few studies, where off-specular polarized neutron scattering has played an important role to understand the magnetic properties in macroscopic length scales.

4.3.1 Magnetic domain disorder and its correlation in antiferromagnetically coupled multilayers

As described in chapter 3 the oscillatory interlayer coupling (OIC) in magnetic multilayer consisting of FM/metallic bilayers plays an important role in the application, e.g. giant magnetoresistance (GMR), etc, and the ground state in these OIC systems is an antiferromagnetic (AF) alignment of neighbouring magnetic layers. The change in resistivity arises from the spin-dependent scattering of the conduction electrons. The scattering further depends on the magnetic moment alignment, the interfacial disorder, and the details of the magnetic domain structure. The Co/Cu multilayer is one such system which has been studied extensively for understanding depth-dependent magnetic structure and coupling. Thickness-dependent and various other studies on Co/Cu multilayers [53] suggested that a vertically incoherent magnetic domain structure will reduce the GMR by preventing perfect AF alignment in adjacent layers. Using PNR measurements, Borchers *et al* [54] have shown dependence of GMR on the vertical correlation of domains in a weakly coupled Co/Cu multilayer at the coercive field and the study suggested that the loss of GMR is directly correlated to the vanishing of the vertical correlation. Therefore magnetic domain structure and its correlation at different interfaces provide an in-depth understanding as well as a mechanism for improving the heterostructure towards application of such magnetic systems.

Using diffuse unpolarized neutron scattering, Langridge *et al* [34] quantitatively determined the field dependence of the domain distribution in strongly coupled Co/Cu multilayers of 50 bilayer repeats, with Cu spacer thickness corresponding to AF coupling. The nominal structure of the multilayer was [Co(20 Å)/Cu(20 Å)]$_{50}$. Figure 4.8 shows the diffuse neutron scattering data as a function of Q_X and Q_Z from the Co/Cu multilayer [34]. Due to the chemical periodicity of the bilayer in the multilayer, a nuclear Bragg peak is observed at $Q_Z = 0.15$ Å^{-1} in figure 4.8(a) (diffuse data at zero field) and in figure 4.8(b) (diffuse data at an applied magnetic field, H, of 700 Oe). At zero field half order [001/2] a Bragg peak is also observed (figure 4.8(a)), which is due to AF magnetic periodicity in the multilayer where alternate Co layers are coupled antiferromagnetically. The half-order Bragg peak intensity vanishes (figure 4.8(b)) at a higher applied field as the coupling changes from AF to FM. It is evident from figure 4.8 that the nuclear Bragg peak (first order) is sharper than the AF peak (half- order), which is entirely magnetic in origin and diffuse.

Assuming a magnetic domain disorder in the same way as the structural height–height correlation was applied in the Born approximation for chemical roughness by Sinha *et al* [1], Langridge *et al* [34] developed a theoretical framework to fit these

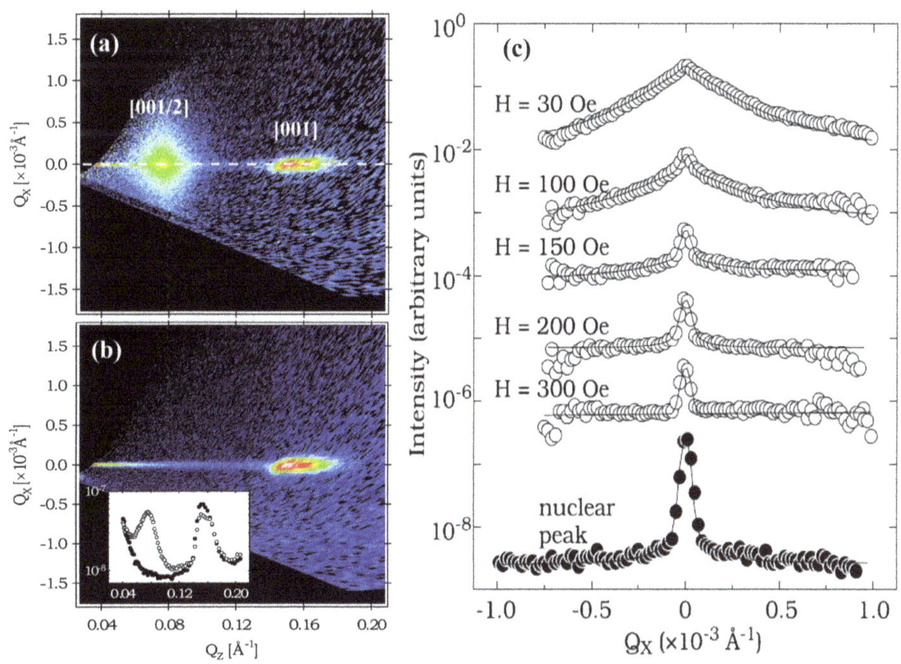

Figure 4.8. (a) The diffuse unpolarized neutron scattering from the [Co(20 Å)/Cu(20 Å)]$_{50}$ multilayer in zero applied field. The intensity along the horizontal dashed line (white) through $Q_X = 0.0$ Å^{-1} represents the specular neutron reflectivity as a function of Q_Z. The intensity centred at $Q_Z = 0.15$ and 0.075 Å^{-1} correspond to the first-order multilayer structural Bragg peak and the AF order, which arises purely from magnetic ordering, respectively. (b) The diffuse neutron scattering measurement in a saturation field of $H = 700$ Oe. The inset shows the specular neutron reflectivity for the low (open symbol) and high (closed symbol) field data. (c) The diffuse magnetic neutron scattering data (open symbol) and fit (line) at the AF peak ($Q_Z = 0.075$ Å^{-1}) as a function of an applied field. Each scan is vertically shifted for clarity. For comparison, the nuclear diffuse data (solid symbols) at saturation ($Q_Z = 0.15$ Å^{-1}) is also plotted. Reproduced with permission from [34]. Copyright 2000 the American Physical Society.

magnetic diffuse scattering data. Figure 4.8(c) shows the field-dependent magnetic diffuse data (open circles) at the AF Bragg peak and the corresponding fit (solid lines) at a different applied fields, suggesting an evolution of magnetic roughness and larger magnetic domain size (in-plane magnetic roughness correlation length) at lower applied fields. For comparing the structure and magnetic roughness they have also fitted the nuclear diffuse data (first-order Bragg peak due to the periodic multilayer) as shown as solid symbols in figure 4.8(c). Thus the field dependence of the diffuse scattering reveals a close relationship between domain disorder, domain size, and interlayer coupling in these technologically important multilayers.

Later, in 2006, Paul *et al* [55] studied the magnetic structure of AF-coupled Co/Cu multilayers by carrying out specular and diffuse polarized neutron scattering measurements as a function of the applied field. They found that the magnetization is distributed in regions with large and small vertically correlated domains. The study also suggested the existence of two length scales for the magnetic domain along the growth direction. The smaller domains with a size of 0.5–1.5 μm, which are saturated at the lower field of ~500 Oe, are most likely found at the top of the multilayer, whereas the larger domains (size ~20 μm) exist beneath the multilayer which remain AF coupled beyond apparent saturation at ~1.5 kOe.

Using specular and off-specular polarized neutron reflectometry, Saerbeck *et al* [56] studied a field and temperature-dependent strong biquadratic coupling in a $Cu_{0.94}Mn_{0.06}$/Co (CuMn/Co) multilayer and investigated the effect of dilute magnetic Mn impurities on the magnetic behaviour of a Co/Cu multilayer system. Analysis of the reflectivity data under DWBA revealed the formation of on average (±30°) canted lateral magnetic domains with a mean width of 0.43 μm extended throughout the multilayer structure. The study also suggested the stabilization of a noncollinear magnetic structure in the Co layer in this CuMn/Co multilayer. The results highlighted the unique capability of PNR in the determination of depth-resolved magnetization profiles in magnetic multilayers. Further, accurate measurements of complicated 2D domain states, coupled with high vertical correlation through the multilayer, were obtained.

The Fe/Cr multilayer is another important system exhibiting GMR properties due to strong AF exchange coupling of adjacent Fe atomic layers. The Fe/Cr multilayer is an excellent model system of an antiferromagnet, in which finite size and surface effects are very well defined as compared to a bulk antiferromagnet. In addition to AF exchange coupling in Fe/Cr multilayers, the depth-dependent spin configuration has been studied both experimentally and theoretically. This system also showed a surface spin-flop transition preceding the 'bulk' phase transition at which magnetization in all layers flips almost perpendicular to the field. It is demonstrated that the magnetization of the alternating Fe layers is twisted through the multilayer stack proving a stable noncollinear configuration and the maximal tilt at the end layers progressively reduces towards the centre of the multilayer [37]. Lauter *et al* [37] studied several Fe/Cr multilayers of different thicknesses and observed the nonuniformly canted state of the spin-flop phase induced by a magnetic field. They observed a Bragg sheet along a line with $Q_z\, d \sim \pi$, where $Q_Z = p_i + p_f$ (p_i and p_f are defined below) is the wave vector transfer component normal to the surface and d (= 76 Å), is the bilayer thickness, in off-specular neutron scattering from a Fe/Cr multilayer suggesting vertically correlated magnetic fluctuation.

Figure 4.9(a) shows the experimental two-dimensional (2D) intensity map scattered from the Fe/Cr multilayer as a function of wave vectors $(p_i - p_f)$ and $(p_i + p_f)$, where p_i and p_f are defined as $p_{i(f)} = \frac{2\pi \sin\theta_{i(f)}}{\lambda}$, where θ_i is the grazing angle of incidence, θ_f is the angle of reflection, λ is the wavelength of the neutron, and (p_i, p_f) are the perpendicular components of the incoming and the outgoing wave vector, respectively. The wave vector transfer is shown in the inset. The logarithmic intensity colour scale is shown on the side. The simulated 2D model for the experimental data (figure 4.9(a)) is shown in figure 4.9(b). The specularly reflected intensity along the line $p_i = p_f$ represents the total thickness oscillations with the first-order Bragg peak at $Q_Z = 0.0826$ Å^{-1} $\sim 2\pi/d$. The absence of measurable diffuse scattering around the first-order Bragg peak suggested negligible interfacial correlated roughness and the intensity around the half-order Bragg peak position at $Q_Z = 0.0413$ Å^{-1} was attributed to a kind of antiferromagnetic ordering across the multilayer stack.

4.3.2 Corrosion-driven structure and magnetic roughness in Ni films

4.3.2.1 Electrochemically controlled corrosion of Ni film

Ni is a prototypical example of a passive metal and has been subject to numerous studies for its corrosion behaviour. Corrosion of Ni in aqueous electrolytes has been the subject of studies for both fundamental queries on the mechanism of metal corrosion and for the high technological importance of Ni and its alloys towards applications. The tendency for pit corrosion of a metal strongly depends on the microstructure and composition. Passivation and pitting of Ni in a Cl^- environment has been studied extensively. However, the depth profile of pit growth along the thickness of the film and the growth of interface morphology, which might play an important role in controlling further growth of the pits, are considerably less understood and studied, mainly due to the lack of methods suitable for the direct structural analysis of the films. Using NR in combination with atomic force microscopy techniques, Singh *et al* [47] studied the electrochemical behaviour of pitting corrosion of a Ni film grown on a Si substrate by sputtering. The pre-passivated Ni film in a chloride-free sulphuric acid solution and subsequently exposure to chloride above the pitting potential showed density degradation along the thickness of the film, suggesting the growth of pits and void networks along the thickness due to corrosion. The atomic force microscopy and off-specular NR measurements suggest that the morphology of the film remains the same after exposure of the film to chloride solution.

Figure 4.10(a) shows a schematic of the Ni film grown on a Si substrate. The electrochemical passivation and pitting corrosion on sputtered Ni surface was carried out by using the AUTOLAB-30 potentiostat/galvanostat. The electrolytes used for controlled corrosion were 0.05 M H_2SO_4 + 0.095 M NaOH without and with 0.05 M NaCl, prepared from high-purity chemicals and Millipore water. The pH of the electrolyte was maintained at 3, making it strongly acidic. The unpolarized NR were performed to investigate the depth profiling of nuclear structure.

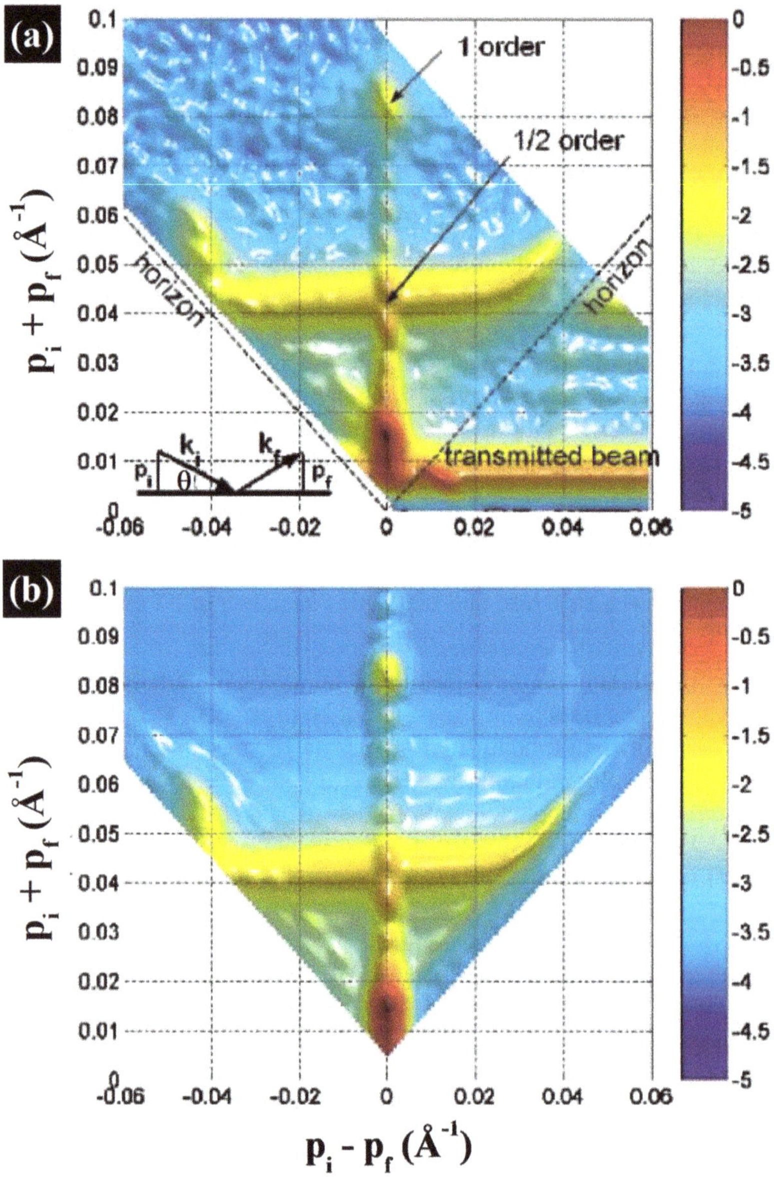

Figure 4.9. (a) Experimental two-dimensional (2D) intensity map scattered from the Fe/Cr multilayer as a function of wave vector ($p_i - p_f$) and ($p_i + p_f$), where p_i and p_f are defined as $p_{i(f)} = \frac{2\pi \sin\theta_{i(f)}}{\lambda}$, with θ and λ the grazing angle of incidence and wavelength of the neutron, the perpendicular components of the incoming and the outgoing wave vector, respectively (as shown in the inset). The logarithmic intensity colour scale is shown on the side. (b) The simulated 2D model for the experimental data in (a). Reproduced with permission from [37]. Copyright 2002 the American Physical Society.

Figures 4.10(b) and (c) show the NR data (symbols) and corresponding fit (solid lines) from the pristine (as-deposited) film (before corrosion) and film after corrosion, respectively. The reflectivity pattern for the film exposed to chloride

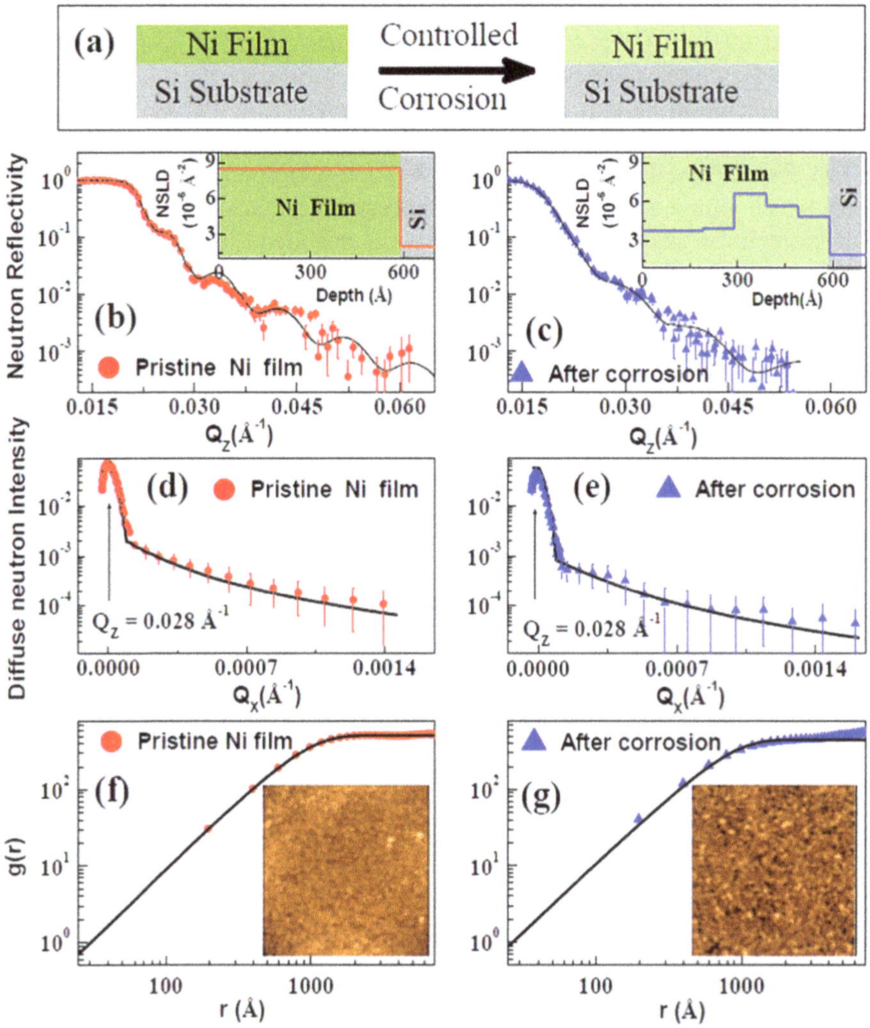

Figure 4.10. (a) Schematic of Ni film on a Si substrate. Specular unpolarized NR data for Ni film before (b) and after (c) corrosion. The inset shows the corresponding nuclear scattering length density (NSLD) depth profiles extracted from the fit to NR data. Off-specular unpolarized neutron scattering data (symbols) and corresponding fit (solid lines) with a specular ridge at $Q_Z = 0.028$ Å^{-1} for (d) pristine and (e) corroded Ni film. The height difference correlation function, $g(r)$ ($=\langle\{Ht(r) - Ht(0)\}^2\rangle$, where Ht is the height on the surface) of the atomic force microscopy data (shown in the inset) of scan area 2.0 μm × 2.0 μm for (f) pristine and (g) corroded Ni film. The morphology parameters from two measurements (off-specular NR and atomic force microscopy) are compared and they suggested good agreement with each other. Reproduced with permission from [47]. Copyright 2009 Elsevier.

(after corrosion) is completely different from that obtained for the pristine film (i.e. before corrosion). The reflectivity pattern for the film after corrosion shows a lower critical angle as well as the modifications in the oscillations (figure 4.10), suggesting an overall loss in the density of the film due to corrosion. The nuclear scattering length density (NSLD) model used to fit the data is shown in the inset of figures 4.10(b) and (c). Diffuse neutron scattering around a perpendicular momentum transfer (Q_Z) of 0.028 $Å^{-1}$ for the pristine and corroded Ni film is shown in figures 4.10(d) and (e), respectively. The morphological parameters, σ (roughness), ξ (in-plane correlation length), and h (Hurst parameter) for the Ni–air interface of pristine Ni film obtained from best fit to diffuse neutron scattering data are 20 Å, 800 Å, and 0.95, respectively. The Hurst parameter $h = 0.95$ corresponds to a fractal dimension of 2.05, which represents a Gaussian two-dimensional surface.

Diffuse neutron scattering measurements on corroded film did not illustrate any drastic change in the morphology as compared to that of pristine film and morphological parameters, $\sigma = 22$ Å, $\xi = 1200$ Å, and $h = 0.90$ were obtained for the corroded film, which again shows a Gaussian two-dimensional surface. A higher correlation length for a corroded Ni film was obtained, which corresponds to the typical pit size.

The morphological parameters obtained from diffuse neutron scattering data for Ni film before and after corrosion were compared with the AFM data from the film. The height (Ht) distributions at the surface of the films obtained from AFM measurements were used to estimate the height difference correlation function (HDFC), $g(r)$, which can be related to a self-affine fractal surface as below:

$$g(r) = < \{Ht(r) - Ht(0)\}^2 > = 2\sigma^2\left[1 - \exp\left(-\frac{r}{\xi}\right)^{2h}\right]. \tag{4.1}$$

Using AFM data the HDCF ($g(r)$) was obtained (symbols) for the pristine (figure 4.10(f)) and corroded (figure 4.10(g)) Ni surface, respectively. The solid lines in figures 4.10(f) and (g) show the fit to $g(r)$ data using self-affine fractal surfaces as given in equation (4.1). The morphological parameters obtained for pristine and corroded Ni film obtained from AFM data closely match with that obtained from diffuse neutron scattering, suggesting a similar kind of morphology for the pristine and corroded films, which are Gaussian and nearly two-dimensional.

4.3.2.2 Separation and correlation of structural and magnetic roughness in a Ni thin film

As mentioned before, diffuse neutron and x-ray scattering have been used extensively for the determination of interface morphology in solid and liquid interfaces. For neutrons, there is another novel advantage where diffuse scattering with polarized neutrons can be used to determine the morphology of a magnetic interface. However, due to the lower brilliance of neutron sources, not many studies have been attempted, even though the magnetic interaction of neutrons with atomic magnetic moments is much easier to comprehend and easily tractable theoretically. Singh *et al* [48] obtained a simple and physically meaningful expression, under the Born approximation (BA), for analyzing diffuse (off-specular) polarized neutron scattering (DPNS) data and studied Ni film on a glass substrate for structural and magnetic morphological correlations. The detailed theoretical analysis of this formalism under BA is given in chapter 2. Under BA, the separation and correlation of chemical and magnetic morphologies were quantified and the results also suggested that the magnetic interface is smoother than the chemical interface. Specular PNR measurements of the sample was carried out to determine the magnetic moment density profile and it suggested a reduction of magnetic moments near the surface.

As mentioned in chapter 2, there are several ways to define magnetic roughness at the interfaces in thin films and Singh *et al* [48] assumed the existence of a structural (chemical) and a separate magnetic boundary buried below it at each interface similar to the reported works [22, 57] and derived a general expression (given in chapter 2) for roughness for the chemical boundary and the magnetic boundary with a correlation term between the chemical and magnetic roughness similar to Osgood *et al* [22]. The inset of figure 4.11(a) represents the actual boundaries, which can be

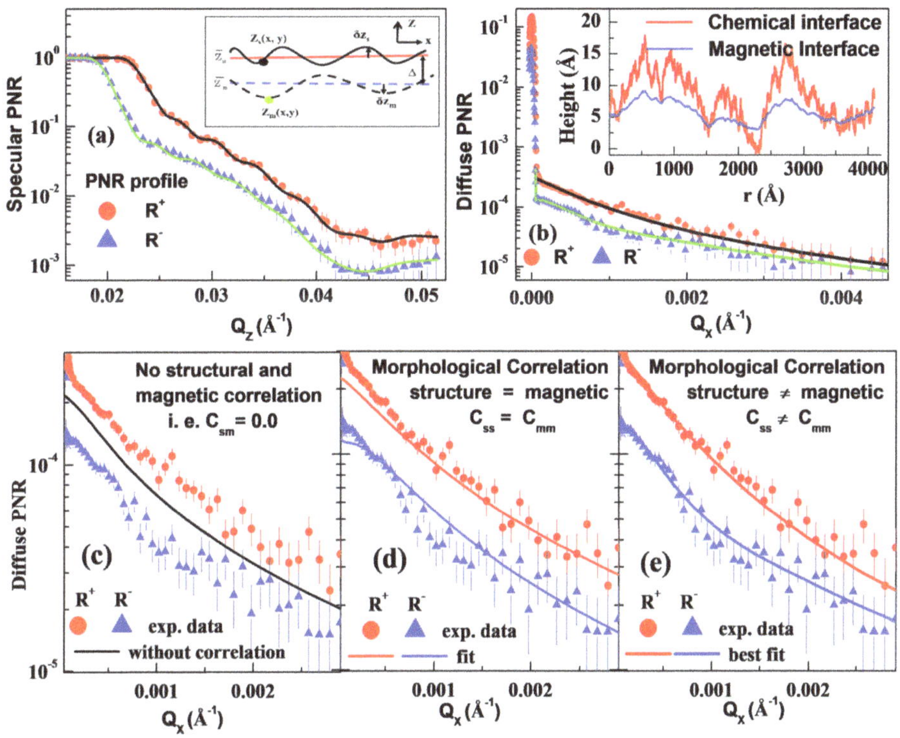

Figure 4.11. (a) Specular polarized neutron reflectometry data from the Ni film on a glass substrate. Closed circles and triangles are the experimental data for spin-up (R^{+}) and spin-down (R^{-}) neutrons as a function of Q_Z. Continuous lines are the best fit. The inset shows a diagram of the structural (chemical) and magnetic boundaries at an interface. This model has been used to calculate the off-specular (diffuse) polarized neutron scattering (DPNS) intensity from a rough interface. (b) Off-specular polarized neutron scattering data (symbol) and corresponding fits (solid lines) from the sample. Closed circles and triangles are the experimental data for spin-up (R^{+}) and spin-down (R^{-}) neutrons, respectively, at $Q_Z = 0.0252$ Å^{-1}. The inset shows a line profile of height distribution at the surface of the film obtained from DPNS data for chemical (structure) and magnetic boundaries. A comparison of fitting to DPNS data considering different morphological parameters for two boundaries (i.e., chemical and magnetic) of the interface are depicted in (c)–(e). The different correlation functions considered are (c) no structure-magnetic correlation, (d) similar structure and magnetic parameters and (e) different structure and magnetic parameters. Reproduced with permission from [48]. Copyright 2009 IOP Publishing.

then considered as really composed of two boundaries, a chemical boundary and a magnetic boundary, each with their average heights, roughnesses, and correlation lengths. Figure 4.11(a) shows the specular PNR data (symbol) and corresponding fit and suggested reduced magnetic moment near the surface [48]. Figure 4.11(b) shows the DPNS data as a function of the in-plane wave transfer vector Q_X for spin-up (R^+) and spin-down (R^-) neutrons measured at $Q_Z = 0.0252$ Å^{-1}.

The DPNS data were collected around the centre on a specular peak beyond the total reflection region but remaining in the weak scattering region where the Born approximation is valid. Over the entire Q_X range, there is a large difference in the intensity between the up and down-polarized beams. The best fit to data suggested different chemical and magnetic boundaries with different morphological parameters (σ, ξ, and h) and the result suggested lower magnetic roughness, larger magnetic correlation length and a higher value of h for a magnetic boundary as compared to the chemical boundary at each interface [48]. The morphological parameters of different boundaries (chemical and magnetic) were further correlated and different cases with fit to DPNS data in a limited Q_X range are shown in figures 4.11(c)–(e). For uncorrelated chemical and magnetic morphological parameters, the DPNS data for both spins should remain the same as shown in figure 4.11(c) by solid black line. But the data suggest that the morphological parameters for chemical and magnetic boundary at each interface are correlated. Two other cases were also considered (i) with similar (figure 4.11(d)) and (ii) different (figure 4.11(e)) morphological parameters for both chemical and magnetic boundaries at each interface and the results suggested that the best fit to the DPNS data were obtained for the case of different morphological parameters for chemical and magnetic interfaces that are correlated (figure 4.11(e)).

The DPNS measurements indicated that the magnetic boundary shows a higher in-plane correlation length with a smoother interface (larger Hurst parameter h). The study demonstrated that diffuse polarized neutron reflectometry can be used as a powerful tool to look at magnetic and chemical (structural) boundaries simultaneously. Further, to visually highlight two boundaries (structural and magnetic) at the air–film interface we have generated the fractal surfaces using a midterm displacement method [58]. We have used the morphological parameters (σ, ξ, and h) for two boundaries extracted from DPNS measurements for rendering the two boundaries at the air–film interface [48]. Figure 4.12 shows a two-dimensional plot of the chemical (curve S) and magnetic (curve M) boundaries of the air–film interface. The chemical boundary (curve S) has sharp undulations that are not present in the magnetic boundary. The square grids represent squares with sides of length 2ξ for both plots. It is evident from figure 4.12 that the magnetic boundary is much smoother ($h = 0.83$) than the chemical boundary ($h = 0.52$) and the chemical boundary shows a highly fractal nature with fractal dimension $D = 2.48$, whereas the magnetic boundary is nearly two-dimensional. Also the correlation length of the magnetic boundary is higher than that for the chemical boundary. The inset (i) of figure 4.12 shows a comparison of the line profiles of two boundaries.

The experimental study in this case for the separation of chemical and magnetic roughness highlights the strength of DPNS measurements to unravel magnetic and structural roughness for interfaces in thin magnetic films of interest. It also

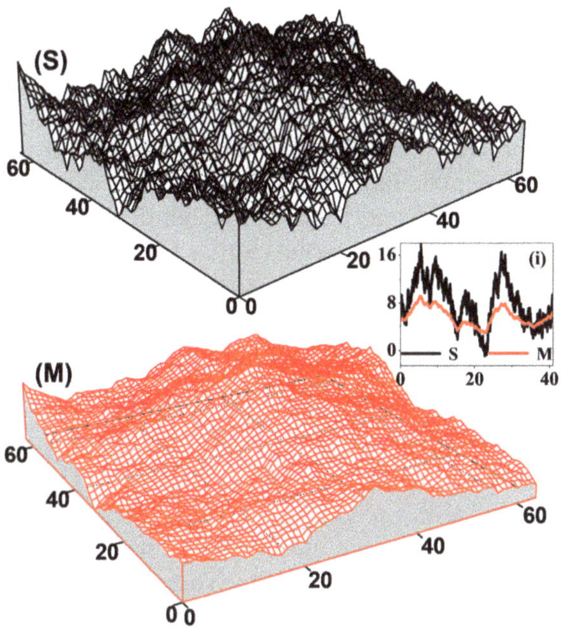

Figure 4.12. Rendition of the structural (S) and magnetic (M) boundary at the air–film interface using the midpoint displacement method for fractal surfaces (see text). The inset (i) shows the one-dimensional height profile (line profile) for two boundaries, structural (S) and magnetic (M).

demonstrates that DPNS can be used as a powerful tool to look at magnetic and chemical interfaces simultaneously. Such experiments are rare and it is presented here for encouraging future studies in this direction.

4.3.3 Helical magnetic structure and evolution of magnetic domains in rare-earth-based multilayers

As discussed in chapter 3 (section 3.3.2), compensated rare earth (RE)–transition metal (TM) alloys and heterostructures, where the RE and TM moments are aligned antiparallel due to the strong AF interaction and the net moment tends to zero, are potential candidate materials for realizing memory devices with higher speed and density. The temperature at which the net moments tend to zero is called compensation temperature (T_{comp}) and the evolution of magnetic structure across T_{comp} is very important for understanding the emerging behaviour shown by these systems as well as improving their technological application. Here we will discuss a few RE/TM-based heterostructures, where diffuse neutron scattering has played an important role in understanding the macroscopic magnetic properties offered by these systems.

4.3.3.1 Evolution of magnetism across the compensation temperature of the Gd/Co multilayer

In chapter 3 (section 3.3.2.1) we discussed the temperature-dependent magnetic structure of the Gd/Co multilayer and correlated the macroscopic magnetic and magnetotransport properties with a microscopic magnetic structure of the multilayer, which showed a helical magnetic structure. However, to understand the antisymmetric magnetoresistance shown by the multilayer at T_{comp}, an investigation of the evolution of the magnetic domains across T_{comp} was necessary. The DPNS measurements with polarization analysis of the scattered neutrons were carried out to understand the evolution magnetic domain and its correlation across T_{comp} (~125 K for one of the Gd/Co multilayers, although the T_{comp} depends on both the thickness and composition of the materials) for the Gd/Co multilayers [50–52]. Spin-dependent specular PNR confirmed an AF coupling at the interface between the Gd and Co layers as well as the formation of a twisted helical magnetic structure with planar 2π domain walls (DWs), which were strongly dependent on the cooling (applied) field [50, 52]. Spin-dependent DPNS with spin analysis measurements revealed the development of magnetic inhomogeneities at the central part (thickness ~25 Å) of each Gd layer in the multilayer for which magnetization is perpendicular to the applied (in-plane) field (or neutron polarization) [50]. The study also suggested that the development of a magnetic domain with a vertically correlated magnetic roughness is contributing toward an antisymmetric magnetoresistance at T_{comp} in the system and the magnetic roughnesses, which are vertically correlated, result in a Bragg sheet observed in the spin-flip channel of DPNS with spin analysis data [50].

Figure 4.13(a) shows the DPNS with spin analysis (R^{++} and R^{+-}) data (Q_X–Q_Z intensity map) from Gd/Co multilayers at 5, 125, and 200 K in an applied in-plane magnetic field of 500 Oe. The intensity as a function of Q_Z at $Q_X = 0.0$ Å^{-1} represents the specular PNR, which provides a depth-dependent magnetic structure and has been discussed in chapter 3. The Q_X–Q_Z intensity map for NSF R^{++} data did not show any diffuse signals at different temperatures. However, strong diffuse signals in the form of a Bragg sheet, i.e. intensity as a function of Q_X at Bragg positions, for the SF (R^{+-}) mode at 125 K were observed, which disappeared at high (200 K) as well as at low (5 K) temperatures suggesting a magnetic origin. Bragg sheets in the R^{+-} intensity map at 125 K (T_{comp}), suggested the development of magnetic inhomogeneities at vertically correlated interfaces. The DPNS with spin analysis data (Q_X–Q_Z intensity map) can be calculated by the supermatrix method [37, 59–61] within the framework of the distorted-wave Born approximation (DWBA). The differential diffuse scattering cross-section is given by [61]

$$\frac{d\sigma}{d\Omega} = \left(\frac{Ak_0^4}{16\pi^2}\right)\left|\sum_{n=1}^{N} C_n\right|^2 S(Q_X), \tag{4.2}$$

where A is the illuminated surface area, k_0 is the wave vector of the incident neutrons, and C_n are the matrix elements given in [61]. For Gaussian-like lateral roughness and vertically correlated interfaces the structure factor, $S(Q_X)$, is defined as [61]

$$S(Q_x) = \frac{\sigma_{\mathrm{m}}^2\xi}{\sqrt{2\pi}}\exp\left(-0.5\frac{Q_x^2}{\xi^2}\right), \tag{4.3}$$

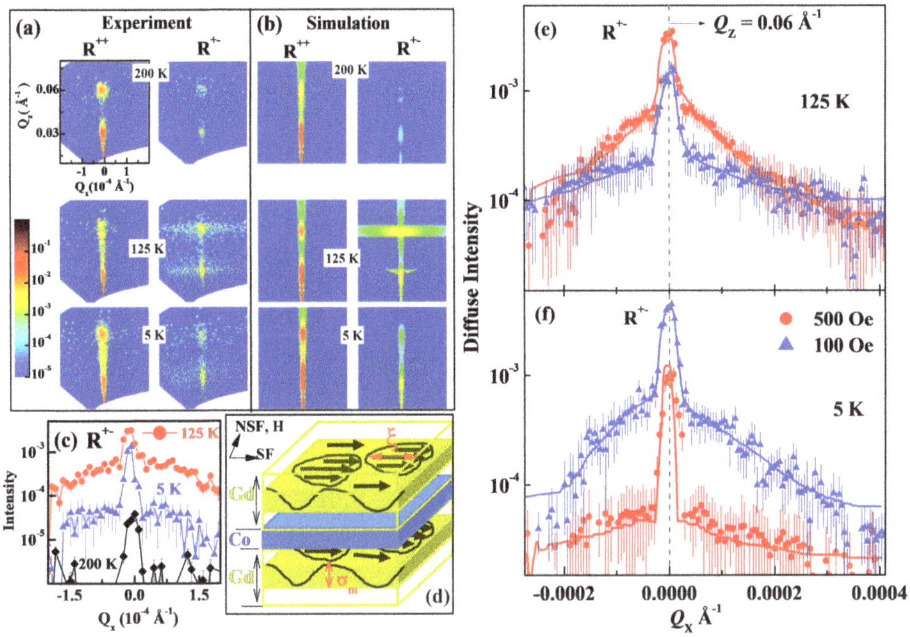

Figure 4.13. (a) Diffuse PNR data (Q_X–Q_Z intensity map) from the Gd/Co multilayer at different temperatures across its compensation temperature (T_{comp} ~ 125 K) in non-spin-flip (NSF, R^{++}) and spin-flip (SF, R^{+-}) modes under an applied in-plane magnetic field (H) of 500 Oe. (b) The corresponding simulated profiles at different temperatures. (c) SF (R^{+-}) data around $Q_Z \sim 0.06$ Å^{-1} at different temperatures. (d) Schematic of spin alignment of the Gd layer in a bilayer, contributing to the Bragg sheet (intensity as a function of Q_X at Bragg peak ($Q_Z \sim 0.06$ Å^{-1})) in R^{+-} intensity at 125 K. Diffuse PNR data (symbols) and corresponding fits (solid lines) for the multilayer at two different in-plane applied fields (500 and 100 Oe) at (e) 125 K and (f) 5 K. Reproduced with permission from [50, 52]. Copyright 2019, 2020 Elsevier.

where σ_m and ξ are magnetic roughness and average lateral correlation length (magnetic domain size in the lateral direction) at the interface. The magnetic roughness (σ_m) can be defined as the root mean square fluctuation of the in-plane projections of magnetization of different domains onto the x-direction (perpendicular to the in-plane applied field). The simulated DPNS with spin analysis (R^{++} and R^{+-}) data (Q_X–Q_Z intensity map) from Gd/Co multilayer at different temperatures, shown in figure 4.13(b) are generated using the superfit routine [59], which uses a supermatrix method [59–61] and the above-mentioned structure factor. In addition, all the parameters obtained from specular PNR were kept fixed and only the magnetic domain distribution at interfaces (magnetic roughness and related correlation length) was varied.

The variations of the SF DPNS intensity (scattered) at the second Bragg peak ($Q_Z \sim 0.06$ Å^{-1}) as a function of Q_X for different temperatures are also compared in figure 4.13(c), which suggests that the scattering at 125 K is purely magnetic in origin. The analysis of DPNS intensity (figures 4.13(a) and (b)) suggested that the Bragg sheets in the SF DPNS intensity map at 125 K are well described by an in-plane correlation length (magnetic domains) ($\sim\xi$) of 0.17 μm at the central part (thickness ~25–30 Å with an rms magnetic roughness (σ_m) ~ 9 Å) in each Gd layer in the multilayer, for which the magnetic moment is aligned perpendicular (in-plane) to the magnetic field (H = 500 Oe), as shown in figure 4.13(d) for a bilayer. The study also suggested a five-fold increase in σ_m for these interfaces at 125 K [50], as compared to that of 200 and 300 K, which are vertically correlated. The growth of magnetic inhomogeneities with magnetization direction perpendicular (in-plane) to the applied field in a part (central 25–30 Å) of the Gd layer (figure 4.13(d)) at 125 K was attributed to the antisymmetric magnetoresistance shown by the multilayer at this temperature [50]. In addition, field-dependent DPNS measurements from the same multilayer demonstrated the evolution of magnetic inhomogeneities (an increase in the magnetic roughness) and magnetic domains of submicron size at different temperatures (125 K for 500 Oe and both 125 and 5 K for 100 Oe), which are highly correlated along the thickness of the multilayer [52]. Figures 4.13(e) and (f) show the one-dimensional (intensity profile as a function of Q_X at Bragg peak profile, i.e.,an intensity along a horizontal line at Bragg peak position in figure 4.13(a)) field-dependent comparison of DPNS data in SF mode (symbols) with simulated profiles (solid lines) as a function of Q_X at the second Bragg peak (Q_Z = 0.06 Å^{-1}) at temperatures 125 and 5 K, respectively.

At 125 K the DPNS SF intensity is higher near Q_X = 0.0 Å^{-1} in an applied field of 500 Oe as compared to that at 100 Oe and it is just the opposite at 5 K. The reduced off-specular SF intensity at 5 K for 500 Oe field suggests a smaller in-plane correlation length. Thus the analysis of DPNS data suggests a distinctively different in-plane correlation length and roughness for the vertically correlated central part of each Gd layer in the two fields. The morphological parameters, ξ (σ_m) of ~0.17 μm (9 Å) and 0.09 μm (5 Å) at 125 and 5 K, respectively, were obtained on cooling the multilayer in a field of 500 Oe [50]. In contrast, at 100 Oe a smaller value of the ξ was obtained at 125 K as compared to that at 5 K [52].

4.3.3.2 Other rare-earth based multilayers

The helical structure of spin in the form of planar DWs is mostly predicted in the theoretical study. However, the helical magnetic structure in rare-earth (RE) based multilayers have been recently reported [49, 50, 55, 62, 63]. Using diffuse polarized neutron scattering, Paul studied the vertically correlated magnetic structures of a RE-based Er/Tb superlattice that leads to long-range magnetic order despite their competing anisotropies along the out-of-plane (Er) and in-plane (Tb) directions [49]. The study suggested the existence of magnetic vortex-like domains, which are associated with a possible magnetic helical ordering within the Er layers.

References

[1] Sinha S K, Sirota E B, Garoff S and Stanley H B 1988 X-ray and neutron scattering from rough surfaces *Phys. Rev.* B **38** 2297–311

[2] Singh S, Swain M and Basu S 2018 Kinetics of interface alloy phase formation at nanometer length scale in ultra-thin films: x-ray and polarized neutron reflectometry *Prog. Mater Sci.* **96** 1–50

[3] Fitzsimmons M R and Majkrzak C 2005 *Modern Techniques for Characterizing Magnetic Materials* (New York: Springer), pp 107–55
[4] Schlomka J P, Tolan M, Schwalowsky L, Seeck O H, Stettner J and Press W 1995 X-ray diffraction from Si/Ge layers: diffuse scattering in the region of total external reflection *Phys. Rev.* B **51** 2311–21
[5] Toperverg B P 2015 Polarized neutron reflectometry of magnetic nanostructures *Phys. Met. Metallogr.* **116** 1337
[6] Singh S, Basu S, Gupta M, Majkrzak C F and Kienzle P A 2010 Growth kinetics of intermetallic alloy phase at the interfaces of a Ni/Al multilayer using polarized neutron and x-ray reflectometry *Phys. Rev.* B **81** 235413
[7] Braslau A, Pershan P S, Swislow G, Ocko B M and Als-Nielsen J 1988 Capillary waves on the surface of simple liquids measured by x-ray reflectivity *Phys. Rev.* A **38** 2457–70
[8] Sanyal M K, Sinha S K, Huang K G and Ocko B M 1991 X-ray-scattering study of capillary-wave fluctuations at a liquid surface *Phys. Rev. Lett.* **66** 628–31
[9] Shpyrko O, Huber P, Grigoriev A, Pershan P, Ocko B, Tostmann H and Deutsch M 2003 X-ray study of the liquid potassium surface: structure and capillary wave excitations *Phys. Rev.* B **67** 115405
[10] Shpyrko O, Fukuto M, Pershan P, Ocko B, Kuzmenko I, Gog T and Deutsch M 2004 Surface layering of liquids: the role of surface tension *Phys. Rev.* B **69** 245423
[11] Russell T P 1990 X-ray and neutron reflectivity for the investigation of polymers *Mater. Sci. Rep.* **5** 171–271
[12] Kjaer K, Als-Neilsen J, Heln C A, Tippmann-Krayer P and Möhwald H 1988 An x-ray scattering study of lipid monolayers at the air–water interface and on solid supports *Thin Solid Films* **159** 17–28
[13] Als-Nielsen J, Jacquemain D, Kjaer K, Leveiller F, Lahav M and Leiserowitz L 1994 Principles and applications of grazing incidence x-ray and neutron scattering from ordered molecular monolayers at the air–water interface *Phys. Rep.* **246** 251–313
[14] Basu J K, Hazra S and Sanyal M K 1999 Growth mechanism of Langmuir–Blodgett films *Phys. Rev. Lett.* **82** 4675–8
[15] Gibaud A, Cowlam N, Vignaud G and Richardson T 1995 Evidence of self-affine rough interfaces in a Langmuir–Blodgett film from x-ray reflectometry *Phys. Rev. Lett.* **74** 3205–8
[16] Singh S, Singh A, Fitzsimmons M R, Samanta S, Prajapat C L, Basu S and Aswal D K 2014 Structural and magnetic depth profiling and their correlation in self-assembled Co and Fe based phthalocyanine thin films *J. Phys. Chem.* C **118** 4072–7
[17] Singh S, Chien T-Y, Guest J R and Fitzsimmons M R 2012 Correlation between surface rumpling and structural phase transformation in $SrTiO_3$ *Phys. Rev.* B **85** 115450
[18] Singh S, Basu S, Bhatt P and Poswal A K 2009 Kinetics of alloy formation at the interfaces in a Ni–Ti multilayer: x-ray and neutron reflectometry study *Phys. Rev.* B **79** 195435
[19] Singh S, Basu S, Bhattacharya D and Poswal A K 2010 Physical and magnetic roughness at metal–semiconductor interface using x-ray and neutron reflectometry *J. Appl. Phys.* **107** 123903
[20] Hannon J P, Trammell G T, Blume M and Gibbs D 1988 X-ray resonance exchange scattering *Phys. Rev. Lett.* **61** 1245–8
[21] Nelson C S, Srajer G, Lang J C, Venkataraman C T, Sinha S K, Hashizume H, Ishimatsu N and Hosoito N 1999 Charge-magnetic roughness correlations in an Fe/Gd multilayer *Phys. Rev.* B **60** 12234–8

[22] Osgood Iii R M, Sinha S K, Freeland J W, Idzerda Y U and Bader S D 1999 X-ray scattering from magnetically and structurally rough surfaces *J. Magn. Magn. Mater.* **198–199** 698–702
[23] Lee D R, Sinha S K, Nelson C S, Lang J C, Venkataraman C T, Srajer G and Osgood R M 2003 X-ray resonant magnetic scattering from structurally and magnetically rough interfaces in multilayered systems. II. Diffuse scattering *Phys. Rev.* B **68** 224410
[24] Basu J K and Sanyal M K 1997 Capillary waves in Langmuir–Blodgett interfaces and formation of confined CdS layers *Phys. Rev. Lett.* **79** 4617–20
[25] Scott J C 2004 Is there an immortal memory? *Science* **304** 62–3
[26] Xue J, Rand B P, Uchida S and Forrest S R 2005 A hybrid planar–mixed molecular heterojunction photovoltaic cell *Adv. Mater.* **17** 66–71
[27] Chen L, Hu Z, Zhao A, Wang B, Luo Y, Yang J and Hou J G 2007 Mechanism for negative differential resistance in molecular electronic devices: local orbital symmetry matching *Phys. Rev. Lett.* **99** 146803
[28] Gredig T, Colesniuc C N, Crooker S A and Schuller I K 2012 Substrate-controlled ferromagnetism in iron phthalocyanine films due to one-dimensional iron chains *Phys. Rev.* B **86** 014409
[29] Kumar A, Singh A, Samanta S, Debnath A K, Aswal D K and Gupta S K 2011 Metallic-like conduction in Co–phthalocyanine/Fe–phthalocyanine composite films grown on sapphire substrates *Appl. Phys. Lett.* **99** 112102
[30] Lytle F W 1964 X-ray diffractometry of low-temperature phase transformations in strontium titanate *J. Appl. Phys.* **35** 2212–5
[31] Singh S, Freeland J W, Fitzsimmons M R, Jeen H and Biswas A 2014 Electronic and magnetic properties of manganite thin films with different compositions and its correlation with transport properties: an x-ray resonant magnetic scattering study *J. Appl. Phys.* **116** 222205
[32] Singh S, Freeland J W, Fitzsimmons M R, Jeen H and Biswas A 2016 Composition dependence of charge and magnetic length scales in mixed valence manganite thin films *Sci. Rep.* **6** 29632
[33] Ankner J F and Felcher G P 1999 Polarized-neutron reflectometry *J. Magn. Magn. Mater.* **200** 741–54
[34] Langridge S, Schmalian J, Marrows C H, Dekadjevi D T and Hickey B J 2000 Quantification of magnetic domain disorder and correlations in antiferromagnetically coupled multilayers by neutron reflectometry *Phys. Rev. Lett.* **85** 4964–7
[35] Toperverg B P 2001 Specular reflection and off-specular scattering of polarized neutrons *Physica* B **297** 160–8
[36] Temst K, Van Bael M J and Fritzsche H 2001 Application of off-specular polarized neutron reflectometry to measurements on an array of mesoscopic ferromagnetic disks *Appl. Phys. Lett.* **79** 991–3
[37] Lauter-Pasyuk V, Lauter H J, Toperverg B P, Romashev L and Ustinov V 2002 Transverse and lateral structure of the spin-flop phase in Fe/Cr antiferromagnetic superlattices *Phys. Rev. Lett.* **89** 167203
[38] Paul A, Kentzinger E, Rücker U, Bürgler D E and Grünberg P 2004 Sequence, symmetry, and magnetic fluctuations of the magnetization reversal in exchange-biased multilayers *Phys. Rev.* B **70** 224410
[39] Gierlings M, Prandolini M J, Fritzsche H, Gruyters M and Riegel D 2005 A PNR study of the off-specular scattering during the asymmetric magnetization reversal in an exchange-biased Co/CoO multilayer *Physica* B **356** 36–40

[40] Paul A, Kentzinger E, Rücker U, Bürgler D E and Grünberg P 2005 Polarized neutron scattering from polycrystalline, exchange-biased magnetic multilayers *Physica* B **356** 26–30

[41] Rücker U, Kentzinger E, Toperverg B, Ott F and Brückel T 2002 Layer-by-layer magnetometry of polarizing supermirrors *Appl. Phys.* A **74** s607–9

[42] Kozhevnikov S V, Ott F, Kentzinger E and Paul A 2007 Enhanced off-specular scattering in magnetic neutron waveguides *Physica* B **397** 68–70

[43] Ott F, Humbert P, Fermon C and Menelle A 2001 Dynamical calculation of neutron scattering on a magnetic grating at grazing incidence *Physica* B **297** 189–93

[44] Theis-Bröhl K, Toperverg B P, Leiner V, Westphalen A, Zabel H, McCord J, Rott K and Brückl H 2005 Correlated magnetic reversal in periodic stripe patterns *Phys. Rev.* B **71** 020403

[45] Langridge S *et al* 2006 Controlled magnetic roughness in a multilayer that has been patterned using a nanosphere array *Phys. Rev.* B **74** 014417

[46] Singh S and Basu S 2006 Microscopic characterization of corrosion morphology: a study in specular and diffuse neutron reflectivity *Surf. Sci.* **600** 493–6

[47] Singh S, Basu S, Poswal A K, Tokas R B and Ghosh S K 2009 Electrochemically controlled pitting corrosion in Ni film: a study of AFM and neutron reflectometry *Corros. Sci.* **51** 575–80

[48] Singh S and Basu S 2009 Separation and correlation of structural and magnetic roughness in a Ni thin film by polarized off-specular neutron reflectometry *J. Phys. Condens. Matter* **21** 055010

[49] Paul A 2016 Stiffness in vortex-like structures due to chirality-domains within a coupled helical rare-earth superlattice *Sci. Rep.* **6** 19315

[50] Singh S *et al* 2019 Antisymmetric magnetoresistance and helical magnetic structure in a compensated Gd/Co multilayer *Phys. Rev.* B **100** 140405

[51] Basha M A *et al* 2020 Helical magnetic structure and exchange bias across the compensation temperature of Gd/Co multilayers *J. Appl. Phys.* **128** 103901

[52] Basha M A, Prajapat C L, Bhatt H, Kumar Y, Gupta M, Kinane C J, Cooper J F K, Langridge S, Basu S and Singh S 2020 Field dependent helical magnetic structure in a compensated Gd/Co multilayer *J. Magn. Magn. Mater.* **516** 167331

[53] Holloway H and Kubinski D J 1998 Magnetoresistive Co/Cu multilayers: hysteresis, polycrystallinity, and irreversible changes on magnetization *J. Appl. Phys.* **83** 2705–14

[54] Borchers J A, Dura J A, Unguris J, Tulchinsky D, Kelley M H, Majkrzak C F, Hsu S Y, Loloee R, Pratt W P and Bass J 1999 Observation of antiparallel magnetic order in weakly coupled Co/Cu multilayers *Phys. Rev. Lett.* **82** 2796–9

[55] Paul A, Kentzinger E, Rücker U, Bürgler D E and Brückel T 2006 Field-dependent magnetic domain structure in antiferromagnetically coupled multilayers by polarized neutron scattering *Phys. Rev.* B **73** 094441

[56] Saerbeck T, Loh N, Lott D, Toperverg B P, Mulders A M, Ali M, Hickey B J, Stampfl A P J, Klose F and Stamps R L 2012 Specular and off-specular polarized neutron reflectometry of canted magnetic domains in loose spin coupled CuMn/Co multilayers *Phys. Rev.* B **85** 014411

[57] Idzerda Y U, Chakarian V and Freeland J W 1999 Quantifying magnetic domain correlations in multilayer films *Phys. Rev. Lett.* **82** 1562–5

[58] Saupe D 1988 Algorithms for random fractals *The Science of Fractal Images* ed M F Barnsley *et al* (New York: Springer), pp 71–136

[59] Rühm A, Toperverg B P and Dosch H 1999 Supermatrix approach to polarized neutron reflectivity from arbitrary spin structures *Phys. Rev.* B **60** 16073–7

[60] Toperverg B P 2015 Polarized neutron reflectometry of magnetic nanostructures *Phys. Met. Metallogr.* **116** 1337–75

[61] Nickel B, Rühm A, Donner W, Major J, Dosch H, Schreyer A, Zabel H and Humblot H 2001 Spin-resolved off-specular neutron scattering maps from magnetic multilayers using a polarized ^{3}He gas spin filter *Rev. Sci. Instrum.* **72** 163–72

[62] Paul A, Mukherjee S, Kreuzpaintner W and Böni P 2014 Exchange-bias-like coupling in a ferrimagnetic Fe/Tb multilayer *Phys. Rev.* B **89** 144415

[63] Fust S, Mukherjee S, Paul N, Stahn J, Kreuzpaintner W, Böni P and Paul A 2016 Realizing topological stability of magnetic helices in exchange-coupled multilayers for all-spin-based system *Sci. Rep.* **6** 33986

Chapter 5

Summary and outlook

Advancements in the interface engineering of heterostructures offer possibilities for fine-tuning electronic and magnetic properties, where minor changes in the interface structure and morphology can induce dramatic changes in transport and magnetic properties. Obtaining emerging properties by designing heterostructures and the subsequent new physics resulting from the interfacial effect has always been an interesting topic for both fundamental research and possible technological applications in the field of spintronics, magnetic devices, etc. It is also noted that the knowledge of the correlation between structure and magnetic properties at the interfaces at the nanoscale length will be required even more in future for studying and developing the functional layer systems (e.g. multiferroic, paleomagnetic, etc). In addition, the development of state-of-the-art thin film growth techniques, e.g. pulsed laser deposition, molecular beam epitaxial (MBE), etc, along with the availability of single crystal substrates allows one to design and deposit heterostructures with excellent epitaxy and preferred orientation that can be tailored. This advancement has provided a way of fabricating atomically sharp and epitaxial interfaces in artificial heterostructures of dissimilar materials that have exhibited novel physical and emerging phenomena, in particular at the interfaces. Therefore a study of depth-dependent structure, magnetic properties, and their correlation is necessary to understand the emerging phenomena occurring at interfaces in heterostructures. A suitable probe of non-destructive character with high depth resolution is a need at the moment to study such phenomena. The efficiency of the simultaneous application of different analytical methods (XRR, PNR, TEM, etc) for engineering and investigation of interface-driven emerging phenomena will be crucial for designing advanced functional materials.

XRR and PNR are two complementary non-destructive techniques that are used routinely to investigate the depth profile of structure and magnetism in a heterostructure with a depth resolution of sub-nanometre length scale. XRR measurements in specular and off-specular modes provide a detailed layer structure (including layer

doi:10.1088/978-0-7503-4695-5ch5

thickness, density, and interface roughness) and in-plane roughness morphology (height–height correlation in the plane). In the case of PNR, the weak interaction of neutrons with matter helps us with the easy interpretation of data in an unambiguous manner. Since neutrons can easily penetrate deeply into most materials without being absorbed, their depth sensitivity makes them a unique technique to explore depth-dependent magnetic structure. PNR with polarization analysis is sensitive not only to the nuclear and magnetization depth profile but also to the depth-dependent vector magnetism of the thin films and heterostructures. The contribution of PNR is significant and extensive in magnetic heterostructure systems.

Interfacial magnetism at the different interfaces in multilayers can perhaps be best investigated using this vector magnetometric technique even for deeply buried layers. Due to the inherent sensitivity of neutrons to magnetic moments, PNR without polarization analysis can be used to determine the depth profile of magnetization density, while PNR with polarization analysis by measuring all four NSF (R^{++} and R^{--}) and SF (R^{+-} and R^{-+}) reflectivities is used to investigate the direction and magnitude of in-plane magnetization layer-by-layer with sub-nanometre depth resolution for vector magnetometry in thin films and heterostructures. Neutron off-specular scattering with polarized neutrons can probe in-plane components of the magnetic structure and also its correlation in the direction normal to the surface plane. Analysis of off-specular polarized neutron scattering can be used to estimate lateral domain structure (including mean domain size, mean angle between magnetization vectors of domains, and dispersion of these angles) at each interface in magnetic heterostructures for relatively larger in-plane length scale of a few microns. PNR as a function of the momentum vector $\left(Q = \frac{4\pi}{\lambda}\sin\theta\right)$ is measured either by changing the angle of incidence (adopted mostly in continuous neutron sources, i.e. reactor sources, which use monochromatic neutron beam) or by using polychromatic beams with a wavelength range (adopted mostly in pulse sources, namely spallation neutron sources) for using the time of flight (TOF) method.

Polarization-dependent absorption of x-rays at a large-scale facility (the synchrotron sources) from magnetic films provides element-specific electronic and magnetic structure because the absorption edges have characteristic energies for each element. Thus x-ray absorption spectroscopy (XAS) utilizes the energy-dependent absorption of x-rays to obtain information about the elemental composition of the sample. Magnetic properties of transition metals are largely related to the *d* orbitals, which are best probed by *L*-edge absorption. XAS involves the excitation by photon absorption of a core-level electron (*p*-orbital in transition metal) into unoccupied states (*d*-orbital) near or above the Fermi level. The difference in the absorption of x-rays on passing through the material in two different polarizations, i.e. dichroism, provides the magnetic properties of the material. The common practice is to use either circularly or linearly polarized x-rays. Using these two types of polarization has the advantage to investigate the detailed contribution of magnetization from the magnetic atom. X-ray magnetic circular and linear dichroism (XMCD and XMLD) measure the dependence of XAS on the helicity of circularly polarized x-rays by a magnetic material. The subsequent filling of the core hole by an electron with lower binding energy results in either the emission of a fluorescence photon or the radiationless emission of an Auger electron.

Therefore, PNR together with magnetic/non-magnetic x-ray scattering provides an important mean of revealing the structural details near and at the surface and vector magnetization with extraordinary spatial detail well beneath the surface together with tagging of elemental magnetic properties. In this book we have included the recent discoveries of emergent magnetic properties at interfaces in heterostructures studied using PNR as a primary technique. We have illustrated these interface-related magnetic properties with a few selected examples and provided extensive references to studies performed using XRR and PNR. These examples and references highlight how PNR has played a crucial role in investigating the induced ferromagnetism and modulation of magnetism at interfaces, particularly in both oxide and non-oxide-based heterostructures. This includes both antiferromagnetic complex oxides, antiferromagnetic and paramagnetic complex oxides, ferromagnetic and antiferromagnetic (multiferroic) oxides, non-magnetic insulating oxides, ferromagnetic/insulator/superconducting complex oxides, magnetic/semiconductor systems, magnetic/magnetic systems, etc. Emerging phenomena arising at interfaces, which are not so important in the bulk matter, caused due to interface break inversion symmetry, Dzyaloshinskii–Moriya interaction, biquadratic, and chiral interactions with detailed spin texture have also been discussed in this book.

In contrast to conventional bulk magnetic measurements, e.g. vibrating sample magnetometers (VSMs) and superconducting quantum interference device (SQUID) magnetometers which are only sensitive to the macroscopic magnetization of the whole sample including substrate (the contribution from the substrates is a serious concern in thin film heterostructures), PNR is a unique tool to probe the interface magnetism with a depth resolution in the sub-nanometre length scale. One of the important highlights of PNR is that the substrate does not contribute to the magnetic signal, unlike in conventional magnetometry measurements where the diamagnetic signal of the substrate often dominates over the tiny magnetic signal originating from the thin film sample. In addition, PNR has several advantages: (i) it is a non-destructive technique for magnetization measurements, averaged over the entire sample surface (several square centimetres); (ii) magnetic fluctuations and domain correlations can be obtained at an in-plane length scale of ~ nanometres to micrometres; (iii) the technique can be applied easily for *in situ* and quasi-real-time magnetization measurements adapting to various types of environments (applied strain, temperature, ranging from ultra-high vacuum to gas atmospheres, kinetics phenomena during growth); (iv) it can measure the behaviour of specific parts by using an isotopic contrast for neutrons; (v) in addition to the interface sensitivity of PNR, it can also address the effect of many other parameters (roughness, interdiffusion, oxygen vacancy, lattice expansion, etc) on interface magnetism, PNR simultaneously provides depth-dependent chemical and magnetic structural information at nanoscales. However, PNR also has some drawbacks in the requirement for sample size (relatively bigger samples are required due to limited neutron intensity), and the model-dependent information extracted from PNR.

Quantum material heterostructures, including superconductors, and topological materials provide exceptional promise for the advancement of next-generation computers, new energy technologies, and high-precision sensors. Emergent topological states in spin–orbit coupled metals, Weyl semi-metals, axion insulators, and topological Mott insulators are other rapidly growing areas of quantum materials, which require unprecedented insights into the fundamental physics of these materials. Interfaces, because of the easy occurrence of charge transfer, orbital ordering, atomic reconstruction, and intermixing in this region, are particularly advantageous for getting energetically favourable new spin phases and properties. It is worth mentioning some of the recent interesting results on advanced quantum heterostructures such as topological insulators and Weyl semi-metals using PNR. Bilayers of the topological insulator Bi_2Se_3 and insulating FM EuS on a sapphire substrate, grown by MBE, showed the presence of interface ferromagnetism in the Bi_2Se_3 well above 300 K from PNR data. Magnetization depth profile in Weyl semimetal films of Co_2MnGa grown on single crystal MgO substrate by sputtering has been obtained using PNR. We have described these two examples in chapter 3. The future of neutron scattering in thin film heterostructure systems is increasingly revolving around the study of magnetic transition metal oxides, multiferroic oxides, and quantum material-based heterostructures. The structure and dynamics in this variety of systems and their interface magnetism will be critically important for opening pathways to new technologies.

Rapid advancement in state-of-art thin film growth techniques for growing chemically homogeneous heterostructures and high-intensity neutron reflectivity techniques will provide an excellent opportunity to enable real-time studies of interfaces of the heterostructures for technological applications in the field of soft matter, energy, and quantum materials. PNR, in particular, is a unique experimental technique to reveal the magnetic structure of interfaces, which can adapt to a range of complex sample environments for *in situ* measurements. The proposed and new reflectometers at upcoming international facilities such as the Spallation Neutron Source (SNS) at Oakridge National Laboratory, USA, and the European Spallation Source (ESS), Lund, Sweden, are being designed with exceptional features of all kinds of sample environments with the possibility of collecting data rapidly and accurately. High neutron flux at these sources is important to study smaller size (lateral size $\leqslant 1$ mm^2) heterostructures and make time-resolved measurements. Emerging phenomena as a result of the accumulation of spin, or creation of spin texture, in these systems are confined at or within a few unit cells in an interface of the heterostructures that will be studied using these highly intense new neutron reflectometers. Here, we provide a list of neutron reflectometers at various major neutron sources in the world (see table 5.1) for benefit of the readers who plan to use the technique for their thin film samples. It is important to mention that the neutron reflectometry technique is available at reactor or spallation neutron sources all over the world. Usually one needs to apply for beam-time for a specific experiment at a source. It is imperative that one carries out detailed characterization of the sample or samples using laboratory based techniques such as x-ray diffraction, x-ray reflectometry, etc, to save time and effort before experiments at a large facility. It is fair to

Table 5.1. A list of neutron reflectometers available around the world in major neutron sources, which are being used for studying the interfaces. The full compilation of neutron reflectometers and the related data analysis program is available at http://www.reflectometry.net/reflect.htm#Analysis.

Name of Reflectometer	Facility	Link to Facility
Platypus	OPAL, ANSTO, Australia	https://www.ansto.gov.au/
REF	BRR, KFKI, Hungary	https://www.szfki.hu/
MR	CSNS, China	http://english.ihep.cas.cn/csns/
PNR	Dhruva, BARC, India	https://www.barc.gov.in/
MARIA	FRM2, Munich, Germany	https://mlz-garching.de/
MIRA	FRM2, Munich, Germany	https://mlz-garching.de/
N-REX+	FRM2, Munich, Germany	https://mlz-garching.de/
REFSANS	FRM2, Munich, Germany	https://mlz-garching.de/
TREFF	FRM2, Munich, Germany	https://mlz-garching.de/
REF-V	HANARO, South Korea	https://www.kaeri.re.kr/mpse
ADAM	ILL, France	https://www.ill.eu/
D17	ILL, France	https://www.ill.eu/
Figaro	ILL, France	https://www.ill.eu/
Inter	ISIS, RAL, UK	https://www.isis.stfc.ac.uk/
PolRef	ISIS, RAL, UK	https://www.isis.stfc.ac.uk/
OffSpec	ISIS, RAL, UK	https://www.isis.stfc.ac.uk/
SOFIA	J-PARC, Japan	http://j-parc.jp/c/en/
SHARAKU	J-PARC, Japan	http://j-parc.jp/c/en/
ASTERIX	LANSCE, LANL, USA	https://lansce.lanl.gov/
CANDOR	NIST, USA	https://www.nist.gov/ncnr
MAGIK	NIST, USA	https://www.nist.gov/ncnr
Polarized Beam Reflectometer	NIST, USA	https://www.nist.gov/ncnr
NG7 Reflectometer	NIST, USA	https://www.nist.gov/ncnr
Liquids Reflectometer	SNS, ORNL, USA	https://neutrons.ornl.gov/sns
Magnetism Reflectometer	SNS, ORNL, USA	https://neutrons.ornl.gov/sns
AMOR	SINQ, PSI, Switzerland	https://www.psi.ch/en
Morpheus	SINQ, PSI, Switzerland	https://www.psi.ch/en
Narziss	SINQ, PSI, Switzerland	https://www.psi.ch/en

stress that PNR, due to its inherent strength to reveal small magnetism at the interfaces with sub-nanometre depth resolution, has been and will continue to play a unique and major role in this exciting research area.

www.ingramcontent.com/pod-product-compliance
Ingram Content Group UK Ltd.
Pitfield, Milton Keynes, MK11 3LW, UK
UKHW051852100926
8556IPUK00005B/586

9 780750 346931